Hodder Gibson

Scottish Examinatio

HIGHER
MATHS

Objective Tests:
Revision and Practice Questions

**HODDER
GIBSON**
PART OF HACHETTE LIVRE UK

The authors are grateful for the help and cooperation of pupils and staff of the Mathematics Department of Bell Baxter High School, Cupar, Fife in testing the questions.

Graph paper graphic © Winston Daridian/iStockphoto.com

Although every effort has been made to ensure that website addresses are correct at time of going to press, Hodder Gibson cannot be held responsible for the content of any website mentioned in this book. It is sometimes possible to find a relocated web page by typing in the address of the home page for a website in the URL window of your browser.

Hachette's policy is to use papers that are natural, renewable and recyclable products and made from wood grown in sustainable forests. The logging and manufacturing processes are expected to conform to the environmental regulations of the country of origin.

Orders: please contact Bookpoint Ltd, 130 Milton Park, Abingdon, Oxon OX14 4SB. Telephone: (44) 01235 827720. Fax: (44) 01235 400454. Lines are open 9.00 – 5.00, Monday to Saturday, with a 24-hour message answering service. Visit our website at www.hoddereducation.co.uk. Hodder Gibson can be contacted direct on: Tel: 0141 848 1609; Fax: 0141 889 6315; email: hoddergibson@hodder.co.uk

© David Smart and Graeme Smart 2008
First published in 2008 by
Hodder Gibson, an imprint of Hodder Education,
part of Hachette Livre UK,
2a Christie Street
Paisley PA1 1NB

Impression number   5 4 3 2
Year   2012 2011 2010 2009 2008

Cover photo © ScPhotos/Alamy
Illustrations by Fakenham Photosettting
Typeset in 1Stone Serif 10/14pt by Fakenham Photosetting, Norfolk
Printed and bound in Great Britain by Martins the Printers, Berwick-upon-Tweed

A catalogue record for this title is available from the British Library

ISBN-13: 978 0340 96518 4

Contents

How to use this book

The questions in this book have been designed to help you prepare for the examination at Higher level in Mathematics. They are intended to do two things.

The first part of the book will provide you with a way of revising for tests and examinations and help you to pinpoint areas of a syllabus in which you may have weaknesses, or techniques in which you may need further practice. To help you to do this the questions are arranged by topic and are in the order in which many students follow the course.

The indicators of the difficulty levels of the questions which appear in brackets beside the answers will help you to measure your own progress against the level of question which you are able to answer correctly.

The difficulty level indicators range from 1 to 5 with the more difficult questions carrying the higher rating. The measures of difficulty are based on a combination of (a) the number of separate steps which must be taken to reach an answer and therefore the scope for misunderstanding or mistakes which the question presents and (b) the depth of understanding and reasoning which is needed to solve the problem.

Please remember that the notes which precede each section are there to help you understand and cope with the questions in the section. They are not intended as a complete course in Higher Mathematics.

The second part of the book, in Test Sets 1 – 6, aims to give you examination practice in answering this style of question which is increasingly popular with examiners, due to the wide coverage of a syllabus which it allows within the time limits of an examination.

The examining bodies which either use or intend to use this type of question include The Scottish Qualifications Authority. The Test Sets in this second part of the book cover a broad range of the syllabus with each question being worth the equivalent of two marks in an examination. No difficulty level indicators are shown for Test Set questions.

All of the items are designed so that the statement of the problem allows you to decide on the answer before looking at the options A to D rather than trying to fit one of the options to the question. The purposes of the book will be best served if you use this method.

Guessing an answer should be avoided as this will not help to improve your examination prospects.

If you find that you can't tackle a question or that you get it wrong, go back to the appropriate part of your notes or textbook and revise the topic before trying the question again.

The questions are intended to be answerable without the use of a calculator.

You should remember that you still need to practise the more extended questions which will appear in your examination and to revise those parts of the syllabus which do not lend themselves to objective questions. Keep working and practising from your text book or your notes and past papers.

May you have all the examination success your work deserves.

Unit 1

1

Section 1 The straight line

> **In this section you need to remember:**

1 The gradient m of a straight line through two points (x_1, y_1) and (x_2, y_2)

   is given by $\dfrac{y_2 - y_1}{x_2 - x_1}$ when $x_1 \neq x_2$.

2 The gradient of a straight line is the tangent of the angle which the line makes with the positive direction of the x-axis, as shown in the diagrams.

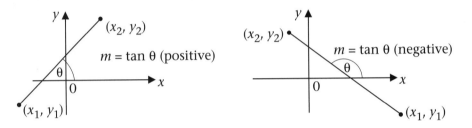

Lines which slope *upwards* to the right have *positive* gradients.

Lines which slope *downwards* to the right have *negative* gradients.

A line parallel to the x-axis has gradient = 0.

A line parallel to the y-axis has an undefined gradient (since x_1 and x_2 are equal and $x_2 - x_1$ would be zero).

3 Two lines are parallel if their gradients are equal ($m_1 = m_2$).

   Two lines are perpendicular if the product of their gradients is -1.
   ($m_1 \times m_2 = -1$).

4 The equation of a line through the origin with gradient m is $y = mx$.

   The equation of a line through the point $(0, c)$ with gradient m is
   $y = mx + c$.

The equation of a line through the point (a, b) with gradient m is $y - b = m(x - a)$.

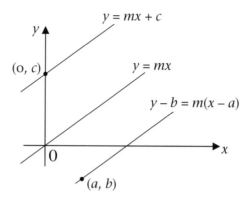

5 The general equation of a straight line is $ax + by = c$ or $ax + by - c = 0$.

The point where two straight lines meet can be found by solving their equations simultaneously.

6 The distance between two points (x_1, y_1) and (x_2, y_2) can be found using the distance formula $d = \sqrt{(x_2 - x_1)^2 + (y_2 - y_1)^2}$.

7 The mid point of a line joining points (x_1, y_1) and (x_2, y_2) is $\{\frac{1}{2}(x_1 + x_2), \frac{1}{2}(y_1 + y_2)\}$.

Revision questions for Unit 1 Section 1:

Question 1
The straight line through points
R (4, 0) and S (−2, 1) has gradient:

A $\dfrac{1}{6}$

B $\dfrac{2}{3}$

C $-\dfrac{1}{2}$

D $-\dfrac{1}{6}$

Question 2
The straight line through points
A (6, 2) and B (4, −1) has gradient:

A $\dfrac{1}{2}$

B $\dfrac{2}{3}$

C $\dfrac{2}{7}$

D $\dfrac{3}{2}$

continued ➢

Revision questions for Unit 1 Section 1: Continued

Question 3

The straight line through points P $(-3, 5)$ and Q $(-1, 2)$ has gradient:

A $\dfrac{2}{3}$

B $\dfrac{3}{2}$

C $-\dfrac{3}{2}$

D $-\dfrac{3}{4}$

Question 4

The line with equation $3y - 7x + 4 = 0$ has gradient:

A $-\dfrac{7}{3}$

B $\dfrac{7}{3}$

C $\dfrac{3}{7}$

D $-\dfrac{4}{3}$

Question 5

The line with equation $3x + 2y = 5$ has gradient:

A $-\dfrac{2}{3}$

B $\dfrac{5}{2}$

C $-\dfrac{3}{2}$

D $\dfrac{3}{2}$

Question 6

A line perpendicular to the line with equation $2x + 3y = 1$ has gradient:

A $-\dfrac{2}{3}$

B $\dfrac{2}{3}$

C $\dfrac{1}{3}$

D $\dfrac{3}{2}$

Question 7

A line perpendicular to the line with equation $2y - 5x = 6$ has gradient:

A $\dfrac{5}{2}$

B $\dfrac{2}{5}$

C $-\dfrac{2}{5}$

D $-\dfrac{5}{2}$

Question 8

A line perpendicular to the y-axis has an equation of the form:

A $y = kx$ (k constant)
B $y = k$ (k constant)
C $x = k$ (k constant)
D $x = 0$

Revision questions for Unit 1 Section 1: Continued

Question 9

A line with equation $y = mx + c$ is perpendicular to a line with equation $y = px + q$ if:

A $p = m$

B $p = -m$

C $p = \dfrac{1}{m}$

D $p = -\dfrac{1}{m}$

Question 10

A line with equation $y = mx + c$ meets the x-axis at an angle whose tangent is:

A $\dfrac{m}{c}$

B $\dfrac{c}{m}$

C m

D unknown without further information

Question 11

A line with gradient -3 which passes through the point $(1, -2)$ has equation:

A $y - 3x = 5$

B $y + 3x = 1$

C $y + 3x = -3$

D $y + 3x = -1$

Question 12

A line with gradient $\dfrac{4}{7}$ which passes through the point $(-2, 4)$ has equation:

A $7y - 4x = 36$

B $7y - 4x = 30$

C $7y - 4x = -30$

D $7y - 4x = 20$

Question 13

In the diagram the line PQ has equation:

A $3x + 4y = 3$

B $4y - 3x = 3$

C $4y - 3x = 12$

D $4y + 3x = 12$

Question 14

In the diagram the line CD has equation:

A $4y - 5x = 5$

B $4y + 5x = 5$

C $4y + 5x = 20$

D $4y - 5x = 20$

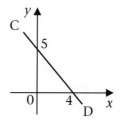

continued ➤

Revision questions for Unit 1 Section 1: Continued

Question 15

In the diagram the line RS has equation:

A $2y + 3x = -3$
B $3y + 2x = -6$
C $2y + 3x = -6$
D $3y + 2x = -3$

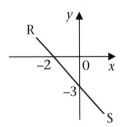

Question 16

In the diagram the line VW has equation:

A $5y - 2x = -10$
B $2y - 5x = -10$
C $5y - 2x = -2$
D $2y - 5x = -2$

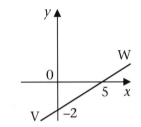

Question 17

The lines with equations
$2x + 7y = 23$ and $2x - y = -1$ meet at the point:

A $(1, 3)$
B $(3, 1)$
C $(1, -3)$
D $(-2, 3)$

Question 18

The lines with equations $2x + y = -1$ and $3x - y = 21$ meet at the point:

A $(22, -45)$
B $(4, 9)$
C $(4, -9)$
D $(22, 45)$

Question 19

In the diagram, the line shown has equation:

A $y - x = 5$
B $y + x = 5$
C $y - x = 1$
D $y + x = 1$

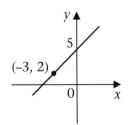

Question 20

In the diagram the line shown has equation:

A $2y - 5x = -18$
B $5y - 2x = 24$
C $2y - 5x = 18$
D $2y + 5x = -2$

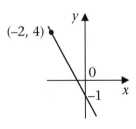

Section 2 Functions and graphs of functions

> **In this section you need to remember:**

1 The function $f(x) = 2x^2 + x - 5$ is a formula which takes any value of a variable x from a given *domain* (all the possible values of x) and finds the corresponding value in the *range* (all the possible values of $f(x)$).

A *composite function* $f(g(x))$ means first apply the function g to the variable x and then apply the function f to the result.

For example if $f(x) = 3x^3$ and $g(x) = x + 6$ then $f(g(x))$ means apply function g to x, giving $(x + 6)$, and then apply function f to $(x + 6)$, giving $3(x + 6)^3$. So $f(g(x)) = 3(x + 6)^3$.

Remember that the **order** of applying the functions is very important as $g(f(x))$ usually gives a *different* end result from $f(g(x))$.

2 The graph of the function $y = h(x)$ crosses the x-axis at points where $h(x) = 0$ (known as the roots), and crosses the y-axis when $x = 0$.

3 The graph of the quadratic function $f(x) = ax^2 + bx + c$ has the shape shown (with respect to the axes) depending on whether a is positive or negative.

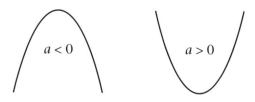

4 The expression $x^2 + 8x + 9$ can be rewritten in the form $(x + 4)^2 - 16 + 9$ or simply $(x + 4)^2 - 7$ by the technique known as *completing the square*. Since the squared term can never be less than zero, the minimum value of the function must be -7. This value arises when $x + 4 = 0$. That is, when $x = -4$.

Similarly the expression $12 - (x + 3)^2$ has a maximum value of 12 when $x = -3$.

continued ➤

5 The graphs of *related functions* may be obtained by sliding the graph of a given function horizontally or vertically or by reflecting it in one of the axes.

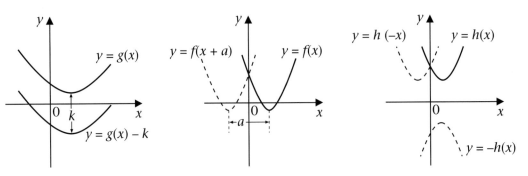

Revision questions for Unit 1 Section 2:

Question 21
If $f(x) = 3x^2$ and $g(x) = x - 4$, then $g(f(x))$ is equal to:

A $9x^2 - 4$
B $3x^2 - 4$
C $3(x - 4)^2$
D $9(x - 4)^2$

Question 22
If $h(x) = 2x - 1$ and $f(x) = \dfrac{2}{x}$ then $h(f(x))$ is equal to:

A $x - 1$

B $\dfrac{4}{x - 1}$

C $\dfrac{2}{2x - 1}$

D $\dfrac{4}{x} - 1$

Question 23
If $f(x) = x^2 + 2$ and $g(x) = x + 1$ then $f(g(3))$ is equal to:

A 12
B 18
C 10
D 36

Question 24
If $h(x) = x^3$ and $f(x) = \cos 2x$ then $f(h(x))$ is equal to:

A $\cos 8x^3$
B $\cos 2x^3$
C $\cos^3 2x$
D $\cos^3 8x^3$

Revision questions for Unit 1 Section 2: Continued

Question 25

Which of the following is most likely to be the equation of the function $f(x)$ in the diagram?

A $f(x) = x^2 + 4x - 3$
B $f(x) = x^2 - 4x + 3$
C $f(x) = x^2 - 2x + 3$
D $f(x) = x^2 - 4x - 3$

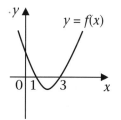

Question 26

Which of the following is most likely to be the equation of the function $g(x)$ in the diagram?

A $g(x) = -x^2 + 2x - 8$
B $g(x) = x^2 - 2x - 8$
C $g(x) = -x^2 + 2x + 8$
D $g(x) = x^2 + 2x - 8$

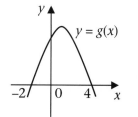

Question 27

Which of the following is most likely to be the equation of the function $h(x)$ in the diagram?

A $h(x) = x^3 - 4x^2 - 4x - 16$
B $h(x) = x^3 - 4x^2 - 4x + 16$
C $h(x) = 16 + 4x + 4x^2 - x^3$
D $h(x) = x^3 + 4x^2 + 4x + 16$

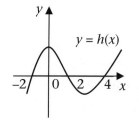

Question 28

Which of the following is most likely to be the equation of the function $k(x)$ in the diagram?

A $k(x) = (x - 1)^2 + 1$
B $k(x) = (x + 1)^2 + 1$
C $k(x) = (x + 1)^2$
D $k(x) = (x - 1)^2$

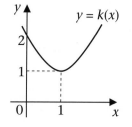

continued ➤

Revision questions for Unit 1 Section 2: Continued

Question 29

When $x^2 - 6x + 5$ is expressed in the form $(x - d)^2 + k$, then d and k have the values:

A $\quad d = 3 \quad k = -4$
B $\quad d = 3 \quad k = 3$
C $\quad d = 6 \quad k = -4$
D $\quad d = -3 \quad k = -4$

Question 30

When $y^2 + 4y - 2$ is expressed in the form $(y + d)^2 + k$, then d and k have the values:

A $\quad d = 2 \quad k = -6$
B $\quad d = 2 \quad k = 6$
C $\quad d = 4 \quad k = -2$
D $\quad d = 4 \quad k = -6$

Question 31

$(x - 4)^2 + 3$ has a turning point (a, k) when a and k have the values:

A $\quad a = 3 \quad k = 4$
B $\quad a = 0 \quad k = 3$
C $\quad a = 4 \quad k = 3$
D $\quad a = 2 \quad k = 3$

Question 32

$(x + 6)^2 - 3$ has a turning point (a, k) when a and k have the values:

A $\quad a = 3 \quad k = -3$
B $\quad a = -6 \quad k = -3$
C $\quad a = -6 \quad k = 3$
D $\quad a = -3 \quad k = -6$

Question 33

If $f(x) = 6x + 2$ and $g(f(x)) = x$, then the function $g(x)$ is equal to:

A $\quad \dfrac{x}{6} - 2$

B $\quad \dfrac{1}{6x + 2}$

C $\quad -6x - 2$

D $\quad \dfrac{x - 2}{6}$

Question 34

If $g(x) = \dfrac{5 - x}{3}$

and $f(g(x)) = x$, then the function $f(x)$ is equal to:

A $\quad 3x - 5$

B $\quad 5 - 3x$

C $\quad \dfrac{3}{5 - x}$

D $\quad \dfrac{-5 + x}{3}$

Revision questions for Unit 1 Section 2: Continued

Question 35

The graph of the function $y = 3^x$ cuts the y-axis at the point:

A (0, 3)
B (1, 0)
C (0, 1)
D (1, 3)

Question 36

The diagram shows the graph of $y = a^x$. The dotted line shows its reflection in the line $y = x$. The equation of the dotted line is:

A $y = \log_x a$
B $y = a^{-x}$
C $y = -a^x$
D $y = \log_a x$

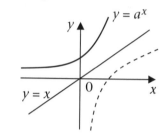

Question 37

The diagram shows the graph of $y = f(x)$. The dotted line shows the graph of:

A $f(x + 4)$
B $f(x) + 4$
C $f(-x)$
D $f(x - 4)$

Question 38

The diagram shows the graph of $y = g(x)$. The dotted line shows the graph of:

A $g(10 - x)$
B $-g(x) + 10$
C $-g(x + 10)$
D $-g(x - 10)$

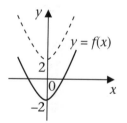

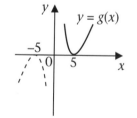

Section 3 Trigonometry: graphs and equations

In this section you need to remember:

1 The graphs of $y = \sin x$ and $y = \cos x$ show maximum values of 1 and minimum values of -1 and have periods of 360 degrees or 2π radians.

2 The graphs of $y = a\sin x$ and $y = a\cos x$ have maximum values of a and minimum values of $-a$.

   The graphs of $y = \sin ax$ and $y = \cos ax$ have periods of $360/a$ degrees or $2\pi/a$ radians.

3 One radian is the angle subtended at the centre of a circle by an arc whose length is equal to the radius. It follows that π radians = 180 degrees.

4 Some exact values of sin, cos and tan are very useful. Make use of these diagrams,

 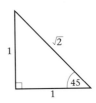

or remember this table of values.

	0° 0 radians	30° $\pi/6$ radians	45° $\pi/4$ radians	60° $\pi/3$ radians	90° $\pi/2$ radians
sin	0	$\dfrac{1}{2}$	$\dfrac{1}{\sqrt{2}}$	$\dfrac{\sqrt{3}}{2}$	1
cos	1	$\dfrac{\sqrt{3}}{2}$	$\dfrac{1}{\sqrt{2}}$	$\dfrac{1}{2}$	0
tan	0	$\dfrac{1}{\sqrt{3}}$	1	$\sqrt{3}$	undefined

5 All the possible values of an angle within a given range must be stated in solving trigonometrical equations.

For example, the equation $4\sin^2 x = 1$ $(0 \le x \le 2\pi)$ gives $\sin^2 x = \dfrac{1}{4}$, so

$\sin x = \pm \dfrac{1}{2}$, so $4\sin^2 x = 1$ has solutions $\dfrac{\pi}{6}$, $\dfrac{5\pi}{6}$, $\dfrac{7\pi}{6}$ and $\dfrac{11\pi}{6}$.

Revision questions for Unit 1 Section 3:

Question 39
The diagram shows the graph of:

A $y = 3 \sin x°$
B $y = 2 \cos 3x°$
C $y = 3 \cos 2x°$
D $y = 3 \sin 2x°$

Question 40
The diagram shows the graph of:

A $y = \cos \dfrac{1}{2}x°$

B $y = \sin 2x°$

C $y = \cos 2x°$

D $y = \sin \dfrac{1}{2}x°$

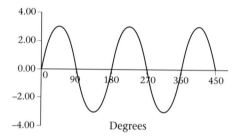

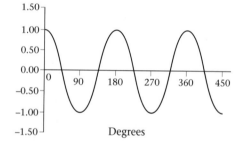

Question 41
The diagram shows the graph of:

A $y = 3 \sin 2x° + 1$
B $y = 2 \sin 3x° + 1$
C $y = 3 \sin 2x° - 1$
D $y = 2 \sin 3x° - 1$

Question 42
The diagram shows the graph of:

A $y = 2 \cos (x - 30)°$
B $y = 2 \cos (x + 30)°$
C $y = \cos 2(x + 30)°$
D $y = \cos 2(x - 30)°$

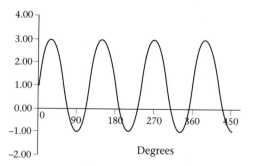

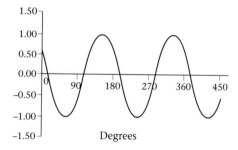

continued ➤

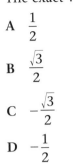

Revision questions for Unit 1 Section 3: Continued

Question 43

The exact value of cos 240° is:

A $\dfrac{1}{2}$

B $\dfrac{\sqrt{3}}{2}$

C $-\dfrac{\sqrt{3}}{2}$

D $-\dfrac{1}{2}$

Question 44

The exact value of tan $\dfrac{2\pi}{3}$ is:

A $-\sqrt{3}$

B $\sqrt{3}$

C $\dfrac{1}{\sqrt{3}}$

D $-\dfrac{1}{\sqrt{3}}$

Question 45

If sin $(x + 30)° = \frac{1}{2}(0 \leq x \leq 360)$,
then x is equal to:

A   0, 120, or 360

B   0 or 120

C   30 or 150

D   120 or 360

Question 46

If cos $^{2}x° = \frac{1}{2}(0 \leq x \leq 360)$ then x has
the value:

A   45 or 315

B   45 or 135

C   45, 135, 225 or 315

D   60, 120, 240 or 300

Question 47

The maximum value of the function
$f(x) = 3 \cos x - 2$ is:

A   3

B   −5

C   1

D   −1

Question 48

The minimum value of the function
$g(x) = 2 - 3 \sin x$ is:

A   2

B   −1

C   5

D   −5

Section 4   Recurrence relations

In this section you need to remember:

1   A linear recurrence relation defines a sequence in which each term is formed from the previous term.

For example, $u_{n+1} = au_n + b$.

If $t_{n+1} = 2t_n + 5$ and $t_0 = 3$, then $t_1 = 2 \times 3 + 5 = 11$;
$t_2 = 2 \times 11 + 5 = 27$; and so on.

2   For a recurrence relation to have a *limit*, the multiplier a must be such that $-1 < a < 1$.

3   If u_n tends to a limit L then $L = \dfrac{b}{(1 - a)}$.

Revision questions for Unit 1 Section 4:

Question 49
If in a recurrence relation
$u_{n+1} = 0.2\, u_n + 4$ and $u_0 = 6$, then u_3
is equal to:

A   1·008
B   5·008
C   5·08
D   5·04

Question 50
If in a recurrence relation
$u_{n+1} = 3u_n - 2$ and $u_0 = -5$, then u_3
is equal to:

A   161
B   53
C   −53
D   −161

Question 51
The sequence defined by the
recurrence relation $u_{n+1} = ku_n + b$
tends to a limit as n tends to infinity if:

A $-1 < b < 1$
B $-1 < k < 1$
C $-1 \le k \le 1$
D $k \le 1$

Question 52
The sequence defined by the
recurrence relation $t_{n+1} = pt_n + q$
tends to a limit as n tends to infinity if:

A $-1 \le p \le 1$
B $-1 < p < 1$
C $-1 < q < 1$
D $-1 \le q \le 1$

continued ➤

Question 53

If a recurrence relation is defined by $u_{n+2} = u_{n+1} + u_n$, and $u_1 = 1$ and $u_2 = 1$, the fifth term of the sequence is:

A 3

B 7

C 8

D 5

Question 54

If the first three terms of a linear recurrence relation $t_{n+1} = mt_n + k$ are 10, 7, and 4 in order, then m and k have the values:

A $m = -3; k = 1$

B $m = 1; k = 3$

C $m = 1; k = -3$

D $m = -2; k = 24$

Question 55

As n tends to infinity, the sequence defined by the recurrence relation $u_{n+1} = 0{\cdot}6u_n + 50$ has a limit of:

A 50

B 125

C 300

D 200

Question 56

As n tends to infinity, the sequence defined by the recurrence relation $t_{n+1} = 0{\cdot}75t_n + 200$ has a limit of:

A 50

B 800

C 200

D 150

Question 57

In the sequence defined by $t_n = 0{\cdot}4\, t_{n-1} + 240$, as n tends to infinity, t_n tends to:

A 600

B 1440

C infinity

D 400

Question 58

In the sequence defined by $p_n = 3p_{n-1} + 60$, as n tends to infinity, p_n tends to:

A infinity

B 120

C 20

D -30

Section 5 Differentiation

In this section you need to remember:

1 The expressions $\dfrac{dy}{dx}$ and $f'(x)$, mean the derivative of y and the derived function of $f(x)$.

2 If $y = ax^n$ then $\dfrac{dy}{dx} = anx^{n-1}$, or if $f(x) = ax^n$ then $f'(x) = anx^{n-1}$

   where a is a constant and n is a rational number.

3 If $f(x) = g(x) + h(x)$ then $f'(x) = g'(x) + h'(x)$.

4 To differentiate a product such as $y = 3x(2x - 1)$, multiply out and differentiate term by term.

   Here, $y = 6x^2 - 3x$, so $\dfrac{dy}{dx} = 12x - 3$.

5 The gradient function of $f(x)$ is $\dfrac{dy}{dx}$ or $f'(x)$.

   The value of the gradient function when $x = k$ gives the *gradient of the curve at a point* whose x coordinate is k.

6 To find the equation of a tangent to a curve at the point (a, b), first differentiate the equation of the curve and substitute a for x. This gives the gradient m of the tangent. Then use the equation $y - b = m(x - a)$ for a straight line.

   For example the curve $y = 3x^2 + 5x - 2$ passes through the point $(1, 6)$.

   The gradient of the tangent to the curve at this point is the value of $\dfrac{dy}{dx}$ at $x = 1$.

   Here $\dfrac{dy}{dx} = 6x + 5$, so the gradient of the tangent is $6 + 5 = 11$.

   The equation of the tangent is then $y - 6 = 11(x - 1)$.

continued ➤

7 A function $g(x)$ is increasing for values of x which make $g'(x)$ positive, decreasing for values of x which make $g'(x)$ negative, and stationary for values of x which make $g'(x) = 0$.

Revision questions for Unit 1 Section 5:

Question 59

If $y = 3x^2 + 5$, then $\dfrac{dy}{dx}$ equals:

A $6x^2$

B $6x + 5$

C $6x$

D $x^3 + 5x$

Question 60

If $y = 6 - 2x^2 + 5x^4$ then $\dfrac{dy}{dx}$ equals:

A $6x - \dfrac{2}{3}x^3 + x^5$

B $-4x + 20x^3$

C $-2x + 4x^3$

D $6 - 4x + 20x^3$

Question 61

If $f(x) = 5x^2 - 2\sqrt{x}$ then $f'(x)$ is equal to:

A $10x - x^{-\frac{1}{2}}$

B $10x + x^{-\frac{1}{2}}$

C $10x - 2x^{-\frac{1}{2}}$

D $10x + 2x^{\frac{1}{2}}$

Question 62

If $f(x) = \dfrac{1}{\sqrt[3]{x^5}}$ then $f'(x)$ is equal to:

A $\dfrac{-5}{3\sqrt[3]{x^8}}$

B $\dfrac{1}{\sqrt[3]{5x^4}}$

C $\dfrac{-5}{3\sqrt{x^3}}$

D $\dfrac{-5}{3\sqrt[3]{x^{\frac{2}{3}}}}$

Question 63

The derivative with respect to x of $(2x - 1)(x - 5)$ is:

A $4x - 9$

B $4x - 11$

C 2

D $2x - 11$

Question 64

The derivative with respect to x of $\sqrt{x}(x + 1)$ is:

A $\dfrac{1}{2}\sqrt{x}\left(3 - \dfrac{1}{x}\right)$

B $\sqrt{x}\left(3 + \dfrac{1}{x}\right)$

C $\sqrt{x}\left(\dfrac{3}{2} + \dfrac{1}{x}\right)$

D $\dfrac{1}{2}\sqrt{x}\left(3 + \dfrac{1}{x}\right)$

Revision questions for Unit 1 Section 5: Continued

Question 65
The gradient of the tangent to the curve $y = 4x^3 - 5x^2 + x - 7$ at the point $(2, 7)$ is:

A   22

B   29

C   1127

D   7

Question 66
At the point where $x = 4$, the tangent to the curve with equation $y = x + \dfrac{1}{x}$ has gradient:

A $\dfrac{1}{2}$

B $4\dfrac{1}{4}$

C $\dfrac{15}{16}$

D $\dfrac{17}{16}$

Question 67
The function $g(x) = 3x^2 + 2$ is an increasing function if:

A $x < 0$

B $x \le 0$

C $x \ge 0$

D $x > 0$

Question 68
The function $f(x) = 2x^3 - 6$ is a decreasing function for:

A $x = 0$

B   all values of x

C   all values of x except $x = 0$

D   no values of x

Question 69
The curve with equation $y = ax^2 + bx + c$ has a turning point when:

A $x = -\dfrac{b}{2a}$

B $x = 0$

C $x = \dfrac{b}{2a}$

D $x = -\dfrac{2a}{b}$

Question 70
The curve with equation $y = x(x - 2)^2$ has a minimum turning point when x is equal to:

A $-\dfrac{2}{3}$

B   0

C $\dfrac{2}{3}$

D   2

continued ➤

Revision questions for Unit 1 Section 5: Continued

Question 71

For what values of x does this curve show a decreasing function?

A $\ 1 < x < 7$
B $\ x < 4$
C $\ x > 4$
D $\ x \leq 4$

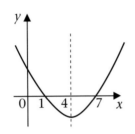

Question 72

For what values of x does this curve show an increasing function?

A $\ x \leq 2$
B $\ x < 2$
C $\ x > 2$
D $\ 0 < x < 4$

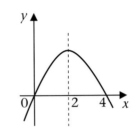

Unit 2

Section 1 Polynomials

> ## In this section you need to remember:
>
> 1 To work out the value of a polynomial for a given value of the variable you can use direct substitution.
>
> For example if $f(x) = 2x^3 - 4x^2 + 5x - 1$,
>
> then $f(3) = 2(3)^3 - 4(3)^2 + 5(3) - 1 = 54 - 36 + 15 - 1 = 32$.
>
> 2 *Synthetic division* is a commonly used way of working out the value of a polynomial function.
>
> For the function $f(x) = 2x^3 - 4x^2 + 5x - 1$ when $x = 3$, it works like this:
>
> $$
> \begin{array}{r|rrrr}
> 3 & 2 & -4 & 5 & -1 \\
>   &   & 6 & 6 & 33 \\
> \hline
>   & 2 & 2 & 11 & \mathbf{32}
> \end{array}
> $$
> so $f(3) = 32$.
>
> (Remember to add the columns and to multiply in the direction of the arrows.)
>
> 3 When a polynomial $f(x)$ is divided by $(x - a)$ the *remainder* is $f(a)$ and can be found by finding the value of $f(x)$ when $x = a$ as shown in 1 or 2 above. The *quotient* can also be read from the bottom line of the synthetic division in 2 above so
>
> $2x^3 - 4x^2 + 5x - 1$ divided by $(x - 3)$ gives $2x^2 + 2x + 11$ and a remainder of 32.
>
> 4 If the remainder on dividing $f(x)$ by $(x - a)$ turns out to be zero, that is if $f(a) = 0$, then $(x - a)$ is a factor of $f(x)$.
>
> Also if $(x - a)$ is a factor of $f(x)$ then $f(a) = 0$.
>
> 5 If $f(x) = k(x - a)(x - b)(x - c)$ then a, b, and c are *roots of the equation* $f(x) = 0$.

6 The equation of the polynomial can be found from its graph by reading off the values of the roots a, b, and c, and then finding the value of k by substituting $(0, k)$ in the equation $y = f(x)$.

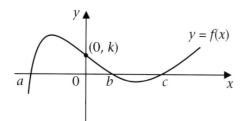

Revision questions for Unit 2 Section 1:

Question 73

If $f(x) = 2x^3 + x^2 - 5x + 1$, then $f(-2)$ is equal to:

A -9

B   3

C -1

D -21

Question 74

If $f(x) = x^4 + 2x^2 - 3x + 3$, then $f(-1)$ is equal to:

A   9

B   12

C   3

D   5

Question 75

When $3x^3 + 2x^2 - 7x + 5$ is divided by $x - 2$ the remainder is:

A   3

B   23

C   17

D   7

Question 76

If $x^4 - x^3 + 2x^2 - px + 4$ divides exactly by $x - 2$, then p has the value:

A -10

B   20

C   18

D   10

Question 77

When $x^4 + 2x^3 - x^2 + 5$ is divided by $x - 3$, the remainder is:

A   131

B   23

C   47

D   150

Question 78

If $x^4 + 2x^3 - kx^2 + 2x - 3$ divides exactly by $x + 3$, then k has the value:

A -2

B $15\frac{1}{3}$

C $3\frac{1}{3}$

D   2

continued ➤

Revision questions for Unit 2 Section 1: Continued

Question 79

When $x^3 - 4x^2 + x + 6$ is divided by $x + 1$, the other factors are:

A $x + 2$ and $x + 3$
B $x - 2$ and $x - 3$
C $x + 6$ and $x - 1$
D $x + 1$ and $x - 6$

Question 80

When $2x^3 + x^2 - 5x + 2$ is divided by $2x - 1$, the other factors are:

A $x + 2$ and $x - 1$
B $x - 2$ and $x + 1$
C $2x - 2$ and $x - 2$
D $2x + 2$ and $x - 1$

Question 81

One root of the equation $x^3 - 21x - 20 = 0$ is 5. The other roots are:

A 1 and 4
B -1 and 4
C -1 and -4
D 1 and -4

Question 82

One root of the equation $4x^3 - 8x^2 - x + 2 = 0$ is $\frac{1}{2}$. The other roots are:

A $\frac{1}{2}$ and -2

B $-\frac{1}{2}$ and 2

C $\frac{1}{2}$ and 2

D $-\frac{1}{2}$ and -2

Question 83

Which of the following is most likely to be the expression for $f(x)$ in the diagram:

A $x^3 - 2x^2 - 5x + 6$

B $\frac{1}{2}(x^3 + 2x^2 - 5x - 6)$

C $x^3 - 2x^2 - 5x + 3$

D $\frac{1}{2}(x^3 - 2x^2 - 5x + 6)$

Question 84

Which of the following is most likely to be the expression for $g(x)$ in the diagram:

A $\frac{1}{3}(x^3 - 7x - 6)$

B $\frac{1}{3}(x^3 + 6x^2 - 7x - 6)$

C $x^3 - 7x - 6$

D $x^3 - 7x + 2$

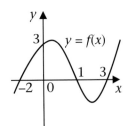

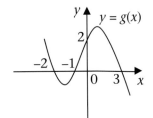

Revision questions for Unit 2 Section 1: Continued

Question 85

If $f(x)$ is a polynomial function, the equation $f(x) = 0$ has a root between a and b if:

A $f(a) > 0$ and $f(b) > 0$

B $f(a) < f(b)$

C $f(a) > 0$ and $f(b) < 0$

D $f(a) > f(b)$

Question 86

If $g(x)$ is a polynomial function such that $g(3) > 0$ and $g(7) < 0$ then the equation $g(k) = 0$ must have a root $x = k$ if:

A $k = 5$

B $3 < k < 7$

C $k = 0$

D $k < 3$ or $k > 7$

Section 2 Quadratic functions

In this section you need to remember:

1 A *quadratic function* is a polynomial in which the highest power of the variable is 2.

Examples are $f(x) = 4x^2 - 3x + 5$, $f(t) = at^2 + bt + c$.

The graph of a quadratic function $f(x) = ax^2 + bx + c$ is a parabola (like this):

$$\underset{a \,<\, 0}{\Bigg(\overset{\text{if}}{}} \quad \text{or} \quad \underset{a \,>\, 0}{\Bigg)\overset{\text{if}}{}}$$

The parabola has an axis of symmetry through its turning point.

2 The roots of a quadratic equation $f(x) = 0$ can be found by factorising the quadratic expression, or by using the quadratic formula

$$x = \frac{-b \pm \sqrt{b^2 - 4ac}}{2a}$$

or by reading them from the graph.

3 *Completing the square* on a quadratic expression by writing it in the form $(x - p)^2 + q$ allows you to write down the equation of the axis of symmetry $x = p$, and to find the maximum or minimum turning point (p, q).

4 For the quadratic expression $ax^2 + bx + c$, the term $(b^2 - 4ac)$ is known as the **discriminant**.

If the discriminant is positive, the roots of the equation $ax^2 + bx + c = 0$ are *real* and are *not equal*.

If the discriminant is equal to zero, the roots of the equation $ax^2 + bx + c = 0$ are *real* and *are equal*.

If the discriminant is negative, the roots of the equation $ax^2 + bx + c = 0$ are *not real*.

5 When finding the points of intersection of a straight line and a parabola, a quadratic equation is formed by solving the equations of the line and the parabola simultaneously.

If in this equation, the discriminant > 0, the line crosses the parabola at two distinct points.

If the discriminant $= 0$, the line is a tangent to the parabola.

If the discriminant < 0, the line does not cross the parabola.

Revision questions for Unit 2 Section 2:

Question 87
The parabola with equation $y = (x - 4)(x + 2)$ has its axis of symmetry on:

A $\quad x = 3$
B $\quad x = 1$
C $\quad y = -1$
D $\quad x = -1$

Question 88
The parabola with equation $y = (2x - 1)(2x + 3)$ has its axis of symmetry on:

A $\quad x = -\dfrac{1}{2}$

B $\quad x = \dfrac{1}{2}$

C $\quad x = 1$

D $\quad x = -1$

Question 89
The parabola with equation $y = x^2 + 5x - 6$ crosses the x-axis when:

A $\quad x = 6$ and $x = -1$
B $\quad x = -6$ and $x = 1$
C $\quad x = 3$ and $x = 2$
D $\quad x = -3$ and $x = 2$

Question 90
The parabola with equation $y = 2x^2 + 9x - 5$ crosses the x-axis when:

A $\quad x = 1$ and $x = -5$

B $\quad x = \dfrac{1}{2}$ and $x = -5$

C $\quad x = -1$ and $x = 2\dfrac{1}{2}$

D $\quad x = -\dfrac{1}{2}$ and $x = 5$

Question 91
The equation $x^2 + 3x - 18 = 0$ has roots:

A $\quad 0$ and -18
B $\quad -6$ and 3
C $\quad 9$ and -9
D $\quad 6$ and -3

Question 92
The equation $6x^2 - 6x - 12 = 0$ has roots:

A $\quad -2$ and 1

B $\quad \dfrac{1}{3}$ and $-\dfrac{1}{6}$

C $\quad 2$ and -1

D $\quad 2$ and $-\dfrac{1}{6}$

continued ➤

Revision questions for Unit 2 Section 2: Continued

Question 93

When $x^2 + 8x + 5$ is expressed in the form $(x + a)^2 - b$:

A $a = 4$ and $b = 5$
B $a = 4$ and $b = 11$
C $a = 4$ and $b = 1$
D $a = 4$ and $b = -11$

Question 94

When $x^2 + 4x + 9$ is expressed in the form $(x + a)^2 + b$:

A $a = 2$ and $b = 7$
B $a = 4$ and $b = 5$
C $a = 2$ and $b = 5$
D $a = -2$ and $b = 13$

Question 95

The expression $x^2 - 6x + 5 \geq 0$ if :

A $x \leq 1$ or $x \geq 5$
B $x \leq 1$ or $x \leq 5$
C $x \geq 1$ and $x \leq 5$
D $x \geq 5$ or $x \geq 1$

Question 96

The expression $x^2 - x - 6 < 0$ if:

A $x > -2$ and $x < 3$
B $x < -3$ and $x > 2$
C $x < -2$ or $x > 3$
D $x > -3$ or $x < 2$

Question 97

The function $6 + x - x^2 > 0$ if:

A $x < -2$ or $x > 3$
B $x > -3$ and $x < 2$
C $x < -3$ or $x > 2$
D $x > -2$ and $x < 3$

Question 98

The function $5x - x^2 \geq 0$ if:

A $x \geq 0$ and $x \leq 5$
B $x \leq \sqrt{5}$
C $x \geq 0$
D $x \leq 0$ or $x \geq 5$

Question 99

The discriminant of the equation $x^2 + 3x + 4 = 0$ is:

A -7
B 9
C -3
D -10

Question 100

The discriminant of the equation $3x^2 + 2x - 7 = 0$ is:

A 86
B 25
C -80
D 88

Question 101

If the equation $x^2 + px + 7 = 0$ has equal roots then:

A $p^2 > 28$
B $p^2 = 28$
C $p^2 = 7$
D $p^2 > 7$

Question 102

If the equation $ax^2 - 3x + k = 0$ has real roots then:

A $-9 - 4ak \geq 0$
B $-3 - 4a^2k^2 \geq 0$
C $-3 - 4ak \geq 0$
D $9 - 4ak \geq 0$

Revision questions for Unit 2 Section 2: Continued

Question 103
If the equation $2x^2 - kx + n = 0$ has no real roots then:

A $\quad -k^2 - 8n < 0$
B $\quad k^2 - 8n > 0$
C $\quad k^2 - 8n = 0$
D $\quad k^2 - 8n < 0$

Question 104
The equation $tx^2 - 2x + 4 = 0$ has real roots if:

A $\quad 4t \leq 1$
B $\quad t \leq 1$
C $\quad 4t \geq 1$
D $\quad t \leq 2$

Question 105
The line $y = x + k$ is a tangent to the curve with equation $y = x^2 + 3x + 5$ if k is:

A $\quad -4$
B $\quad 16$
C $\quad 4$
D $\quad \geq 4$

Question 106
The line $y = 3 - x$ crosses the curve with equation $y = x^2 + x + p$ in two distinct points if p is:

A $\quad < -2$
B $\quad < 4$
C $\quad < 3$
D $\quad < -4$

Question 107
The line $y = 3 + k$ will not cut the curve with equation $y = 2 - x^2$ if k is:

A $\quad > -1$
B $\quad < 2$
C $\quad < -2$
D $\quad < -1$

Question 108
The curve with equation $y = ax^2 + bx + c$ has the shape shown in the diagram if:

A $\quad y > 0$
B $\quad y < 0$
C $\quad a > 0$
D $\quad a < 0$

Section 3 Integration

In this section you need to remember:

1 Integration is the inverse process of differentiation.

If $f(x) = F'(x)$ then

$\int f(x)\, dx$ (read as the integral of $f(x)$ with respect to x) $= F(x) + c$

where c is the constant of integration.

2 The rules of integration are these.

- If $f(x) = ax^n$ then $\int f(x)\, dx = \int ax^n\, dx = \dfrac{ax^{n+1}}{n+1} + c.$

  (Here, a is a constant, $n \neq -1$, and c is the constant of integration.)

- $\int \{f(x) + g(x)\}dx = \int f(x)\, dx + \int g(x)\, dx.$

- When integrating a product or a quotient, first multiply or divide out, and then integrate term by term.

3 An integral of the form $\int f(x)\, dx$ is called an *indefinite* integral.

When limits are given as in $\int_a^b f(x)\, dx$, this becomes a *definite* integral and can be evaluated to give the area under the curve $y = f(x)$ between the limits a and b.

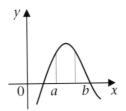

Remember that when evaluating a definite integral, the constant of integration is eliminated.

4 The area between two curves or between a curve and a line is determined by first finding their points of intersection (by solving their equations simultaneously), and then using the x coordinates as the limits of integration.

The area between the curves $y = f(x)$ and $y = g(x)$ between the limits a and b is given by

$$\int_a^b f(x)\, dx - \int_a^b g(x)\, dx = \int_a^b \{f(x) - g(x)\}\, dx.$$

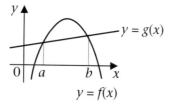

Revision questions for Unit 2 Section 3:

Question 109
The integral $\int x^5\, dx$ is equal to:

A $5x^4 + c$

B $\dfrac{1}{5}x^6 + c$

C $\dfrac{1}{6}x^6 + c$

D $x^6 + c$

Question 110
The integral $\int (5x^4 - x^3 - x)\, dx$ is equal to:

A $x^5 - \dfrac{1}{4}x^4 - 1 + c$

B $5x^5 - x^4 - x^2 + c$

C $x^5 - \dfrac{1}{4}x^4 - \dfrac{1}{2}x^2 + c$

D $\dfrac{1}{4}x^5 - \dfrac{1}{3}x^3 - x^2 + c$

Question 111
The integral $\int (2x^3 - 6x^2 + 1)\, dx$ is equal to:

A $2x^4 - 6x^3 + x + c$

B $\dfrac{1}{2}x^4 - 2x^3 + x + c$

C $3x^2 - 12x + c$

D $\dfrac{2}{3}x^4 - 3x^3 + x + c$

Question 112
The integral $\int 3x^{-6}\, dx$ is equal to:

A $-\dfrac{3}{5}x^{-5} + c$

B $-18x^{-7} + c$

C $-\dfrac{1}{2}x^{-5} + c$

D $\dfrac{3}{5}x^{-5} + c$

continued ➤

Question 113

The integral $\int 3\sqrt{x^4}\, dx$ is equal to:

A $\frac{7}{3}\sqrt[7]{x^3} + c$

B $\frac{4}{3}\sqrt[3]{x} + c$

C $\frac{7}{3}\sqrt[3]{x^7} + c$

D $\frac{3}{7}\sqrt[3]{x^7} + c$

Question 114

The integral with respect to x of

$$\frac{x^2 + 2x - 7}{\sqrt{x}} \text{ is:}$$

A $\dfrac{\frac{1}{3}x^3 + x^2 - 7x}{\frac{2}{3}x^{\frac{3}{2}}} + c$

B $\frac{2}{5}\sqrt{x^5} + \frac{4}{3}\sqrt{x^3} - 14\sqrt{x} + c$

C $\frac{5}{2}\sqrt{x^5} + \frac{3}{2}\sqrt{x^3} - \frac{7}{2}\sqrt{x} + c$

D $\frac{3}{2}\sqrt{x^3} + \frac{1}{\sqrt{x}} + \frac{7}{2}\sqrt{\frac{1}{x^3}} + c$

Question 115

The value of $\int_1^3 (x^2 + 1)dx$ is:

A $28\frac{2}{3}$

B $8\frac{2}{3}$

C $10\frac{2}{3}$

D $13\frac{1}{3}$

Question 116

The value of $\int_{-2}^0 (3 - x + 2x^2)dx$ is:

A $2\frac{2}{3}$

B $16\frac{1}{3}$

C 16

D $13\frac{1}{3}$

Question 117

The value of $\int_1^3 \frac{1}{x^2}\, dx$ is:

A $\frac{26}{27}$

B $1\frac{1}{3}$

C $\frac{80}{81}$

D $\frac{2}{3}$

Question 118

The value of $\int_{-2}^2 (3t - 1)^2 dt$ is equal to:

A 52

B -24

C 48

D 148

Revision questions for Unit 2 Section 3: Continued

Question 119

If the shaded region in the diagram has area 10 units², the value of k is:

A 3

B $\dfrac{4}{3}$

C 7

D 6

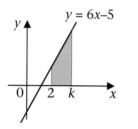

Question 120

The area enclosed between the curve and the straight line in the diagram is:

A $3\dfrac{1}{2}$

B $1\dfrac{1}{2}$

C $2\dfrac{5}{6}$

D $2\dfrac{1}{6}$

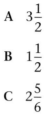

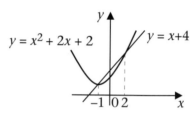

Question 121

The shaded region in the diagram has area:

A 27

B $\dfrac{2}{3}\sqrt{27}$

C 18

D $\sqrt{216}$

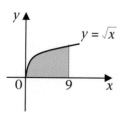

Question 122

The shaded region in the diagram has area:

A $22\dfrac{1}{2}$

B 15

C $10\dfrac{1}{2}$

D $4\dfrac{1}{2}$

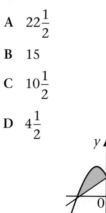

Section 4 3-D Trigonometry and addition formulae

In this section you need to remember:

1   A line is perpendicular to a plane if it makes a right angle with any line on the plane through the point of intersection.

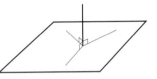

2   The angle between a line and a plane is the angle between the line and its projection on to the plane.

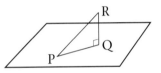

If RQ is perpendicular to the plane then PQ is the projection of PR on to the plane.

The angle between PR and the plane is angle RPQ.

3   The angle between two planes is the angle between any two lines (one on each plane) which meet on the line of intersection and are at right angles to that line.

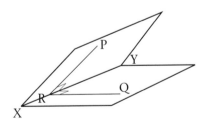

If both PR and QR are at right angles to the line of intersection XY then angle PRQ is the angle between the planes.

4 If tan A is given as $\dfrac{p}{q}$ and angle A is an acute angle, then a right-angled triangle can be drawn and the exact values of other ratios can be found.

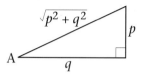

The same method can be used when other ratios are given.

5 The addition formulae are: sin (A + B) = sin A cos B + cos A sin B

sin (A − B) = sin A cos B − cos A sin B

cos (A + B) = cos A cos B − sin A sin B

cos (A − B) = cos A cos B + sin A sin B.

From these, other useful formulae can be derived such as

sin 2A = 2 sin A cos A

cos 2A = cos² A − sin²A

$\qquad$ = 2 cos² A − 1

$\qquad$ = 1 − 2 sin² A.

All of these formulae can be used in solving trigonometric equations and other problems.

For example, in solving the equation cos 2x° − cos x° = 0 for 0 ≤ x ≤ 360, cos 2x° can be written as 2cos² x° − 1.

The left hand side of the equation then becomes 2cos² x° −1 − cos x°.

So 2cos² x° − cos x° − 1 = 0

⇒ (2cos x° + 1)(cos x° − 1) = 0

⇒ 2cos x° +1 = 0, or cos x° − 1 = 0

⇒ cos x° = $-\dfrac{1}{2}$ or cos x⁰ = 1

so x = 120, 240 or x = 0, 360.

Revision questions for Unit 2 Section 4:

Question 123

ABCDEFGH is a cuboid. The angle between the line AG and the plane ABCD is given by:

A ∠GAB
B ∠GAC
C ∠GAD
D ∠GAF

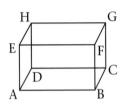

Question 124

PQRSTUVW is a cuboid. The angle between the planes PUVS and PQRS is given by:

A ∠VSR
B ∠VSP
C ∠VSW
D ∠UPR

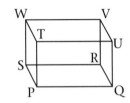

Question 125

PABCD is a square-based pyramid with each of the sloping faces an isosceles triangle. If the length of the base is 6 units and the vertical height is 4 units, the angle between each sloping face and the base is given by:

A $\tan^{-1} \dfrac{3}{4}$

B $\tan^{-1} \dfrac{2}{3}$

C $\tan^{-1} \dfrac{4}{3\sqrt{2}}$

D $\tan^{-1} \dfrac{4}{3}$

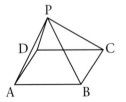

Question 126

PQRSTUVW is a cube with sides of length 5 units. The tangent of the angle between WQ and the shaded face is given by:

A $\dfrac{1}{\sqrt{2}}$

B $\sqrt{2}$

C $\dfrac{1}{2\sqrt{2}}$

D $\dfrac{5}{2\sqrt{5}}$

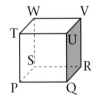

Revision questions for Unit 2 Section 4: Continued

Question 127

If $\cos x° = \dfrac{12}{13}$ and $0 < x < 180$, then the exact value of tan x is:

A $\dfrac{12}{5}$

B $\dfrac{5}{12}$

C $\dfrac{5}{13}$

D $\dfrac{5}{6}$

Question 128

If $\cos α° = \dfrac{20}{29}$ and $0 < α < 180$ then the exact value of sin $α$ is:

A $\dfrac{21}{29}$

B $\dfrac{22}{29}$

C $\dfrac{9}{29}$

D $\dfrac{29}{21}$

Question 129

If $α$ and $β$ are acute angles such that $\sin α = \dfrac{3}{5}$ and $\sin β = \dfrac{8}{17}$, then the exact value of sin $(α + β)$ is:

A $\dfrac{84}{85}$

B $\dfrac{77}{85}$

C $\dfrac{13}{85}$

D $\dfrac{91}{85}$

Question 130

If $θ$ and $φ$ are acute angles such that $\cos θ = \dfrac{12}{13}$ and $\cos φ = \dfrac{8}{10}$, then the exact value of cos $(θ + φ)$ is:

A $\dfrac{126}{130}$

B $-\dfrac{66}{130}$

C $\dfrac{66}{130}$

D $\dfrac{12}{130}$

Question 131

Expressing $15°$ as $(60 − 45)°$ gives the exact value of sin $15°$ as:

A $\dfrac{\sqrt{6} + 2}{2\sqrt{2}}$

B $\dfrac{\sqrt{3} + 1}{2\sqrt{2}}$

C $\dfrac{\sqrt{3} − 1}{2\sqrt{2}}$

D $\dfrac{1 − \sqrt{3}}{2\sqrt{2}}$

Question 132

If OP has gradient 4 and OQ has gradient $\dfrac{1}{4}$ then cos ∠POQ is:

A $\dfrac{8}{\sqrt{17}}$

B $\dfrac{15}{17}$

C $\dfrac{8}{17}$

D $\dfrac{15}{\sqrt{17}}$

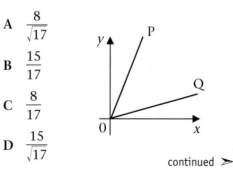

continued ➤

Revision questions for Unit 2 Section 4: Continued

Question 133

If P is an acute angle with $\sin P = \dfrac{12}{13}$, then the exact value of sin 2P is:

A $\dfrac{120}{169}$

B $\dfrac{24}{13}$

C $\dfrac{60}{169}$

D $\dfrac{120}{13}$

Question 134

If θ is an acute angle with $\cos \theta = \dfrac{8}{17}$, then the exact value of cos 2θ is:

A $\dfrac{161}{289}$

B $\dfrac{240}{289}$

C $\dfrac{16}{17}$

D $-\dfrac{161}{289}$

Question 135

The full solution of the equation $\sin 2x° + 2\cos x° = 0$ when $0 \leq x \leq 360$, gives x as:

A   0 or 180

B   90

C   270

D   90 or 270

Question 136

The full solution of the equation $\cos 2\theta° + 5 \cos \theta° + 4 = 0$ when $0 \leq \theta \leq 360$, gives θ as:

A   0 or 360

B   180

C   120 or 240

D   90 or 270

Question 137

The full solution of the equation $\sin 2\alpha - \sin \alpha = 0$ when $0 \leq \alpha \leq 2\pi$ gives α as:

A $0, \dfrac{1}{3}\pi, \pi, \dfrac{5}{3}\pi$ or 2π

B $0, \pi,$ or 2π

C $0, \dfrac{1}{3}\pi,$ or π

D $0, \dfrac{1}{6}\pi, \pi, \dfrac{5}{6}\pi$ or 2π

Question 138

The full solution of the equation $\cos 2\beta + \sin \beta = 0$ when $0 \leq \beta \leq 2\pi$ gives β as:

A $\dfrac{1}{2}\pi, \dfrac{7}{6}\pi, \dfrac{11}{6}\pi$ or π

B $\dfrac{1}{2}\pi,$ or $\dfrac{7}{6}\pi$

C $\dfrac{1}{2}\pi, \dfrac{7}{6}\pi$ or $\dfrac{11}{6}\pi$

D $0, \dfrac{2}{3}\pi, \dfrac{4}{3}\pi$ or 2π

Section 5 The Circle

> **In this section you need to remember:**

1 The work on the circle is based on the formula $d = \sqrt{(x_2 - x_1)^2 + (y_2 - y_1)^2}$ where d is the distance between points (x_1, y_1) and (x_2, y_2).

2 The equation of a circle with its centre at the origin and having radius r is $x^2 + y^2 = r^2$.

3 The equation of a circle with its centre at the point (a, b) and having radius r is $(x - a)^2 + (y - b)^2 = r^2$.

4 The general equation of a circle is $x^2 + y^2 + 2gx + 2fy + c = 0$ (g, f, and c being constants).

5 In the general equation of the circle the centre is the point $(-g, -f)$ and the radius has length $\sqrt{g^2 + f^2 - c}$.

6 To find the points of intersection of a straight line and a circle the equations of the line and the circle have to be solved simultaneously. This involves solving a quadratic equation.

   (a) If the quadratic equation has two roots, the line and the circle meet at two distinct points.

   (b) If the quadratic equation has equal roots, the line is a tangent to the circle.

   (c) If the quadratic equation has no real roots, the line does not meet the circle.

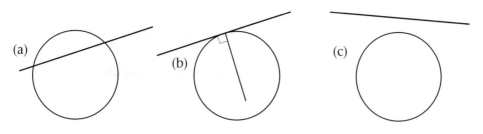

(a)   (b)   (c)

continued ➤

7 The tangent to a circle at a given point is perpendicular to the radius at that point, therefore the product (gradient of radius) $\times$ (gradient of tangent) $= -1$.

For example, if a circle has centre (2, 5) and a tangent to the circle touches it at the point $(-1, 3)$ then the gradient of the radius from (2, 5) to $(-1, 3)$ is

$$\frac{3 - 5}{-1 - 2} = \frac{-2}{-3} = \frac{2}{3}.$$

So the gradient of the tangent is $-\frac{3}{2}$ and the equation of the tangent is

$$y - 3 = -\frac{3}{2}(x - (-1))$$

or $y - 3 = -\frac{3}{2}(x + 1)$.

Revision questions for Unit 2 Section 5:

Question 139
The distance between the points (3, 4) and $(-1, -2)$ is:

A $\sqrt{20}$
B $\sqrt{40}$
C $\sqrt{52}$
D $\sqrt{50}$

Question 140
The distance between the points $(-1, 6)$ and $(2, -4)$ is:

A $\sqrt{101}$
B $\sqrt{66}$
C $\sqrt{89}$
D $\sqrt{109}$

Question 141
If the point $(-2, p)$ lies on the circle with equation $x^2 + y^2 = 36$ then p is equal to:

A $\pm\sqrt{40}$
B $\pm\sqrt{32}$
C $\pm\sqrt{38}$
D $\pm\sqrt{34}$

Question 142
If the point $\big((p + 1), (p + 2)\big)$ lies on the circle with equation $x^2 + y^2 = 41$, then p is equal to:

A 3 or -6
B 4 or 5
C -4 or -5
D -3 or 6

Revision questions for Unit 2 Section 5: Continued

Question 143
If the point $(3, k)$ lies outside the circle with equation $x^2 + y^2 = 16$ then:

A $\quad k < -5$ or $k > 5$
B $\quad -\sqrt{13} < k < \sqrt{13}$
C $\quad -\sqrt{7} < k < \sqrt{7}$
D $\quad k < -\sqrt{7}$ or $k > \sqrt{7}$

Question 144
If the point $(t, -3)$ lies inside the circle with equation $x^2 + y^2 = 49$ then:

A $\quad -\sqrt{40} < t < \sqrt{40}$
B $\quad -\sqrt{58} < t < \sqrt{58}$
C $\quad -\sqrt{46} < t < \sqrt{46}$
D $\quad -\sqrt{40} > t > \sqrt{40}$

Question 145
The circle with centre $(2, 3)$ and radius 6 has equation:

A $\quad x^2 + y^2 + 4x + 6y - 23 = 0$
B $\quad x^2 + y^2 - 4x - 6y - 23 = 0$
C $\quad x^2 + y^2 - 6x - 4y - 23 = 0$
D $\quad x^2 + y^2 - 4x - 6y - 49 = 0$

Question 146
The circle with centre $(-2, 5)$ and radius $\sqrt{29}$ has equation:

A $\quad x^2 + y^2 - 4x + 10y = 0$
B $\quad x^2 + y^2 - 10x + 4y = 0$
C $\quad x^2 + y^2 + 4x - 10y = 0$
D $\quad x^2 + y^2 + 4x - 10y + 29 = 0$

Question 147
The circle with equation $x^2 + y^2 + 14x - 2y + 5 = 0$ has its centre at:

A $\quad (-14, 1)$
B $\quad (1, -7)$
C $\quad (7, -1)$
D $\quad (-7, 1)$

Question 148
The circle with equation $x^2 + y^2 - 8x + 6y + 12 = 0$ has radius:

A $\quad \sqrt{37}$
B $\quad \sqrt{13}$
C $\quad \sqrt{88}$
D $\quad \sqrt{112}$

Question 149
If the point $(8, q)$ lies on the circle with equation $x^2 + y^2 - 8x + 2y - 8 = 0$ then the two possible values of q are:

A $\quad -2$ or 4
B $\quad 2$ or 4
C $\quad -2$ or -4
D $\quad 2$ or -4

Question 150
If the point $(t, 3)$ lies outside the circle with equation $x^2 + y^2 - 2x - 4y - 12 = 0$ then the range of values of t is:

A $\quad t > -5$ and $t < 3$
B $\quad t < 5$ and $t > -3$
C $\quad t < -5$ and $t > 3$
D $\quad t < -3$ and $t > 5$

continued ➤

Revision questions for Unit 2 Section 5: Continued

Question 151

The circle with equation
$x^2 + y^2 + 2x + 2y + 1 = 0$ and the
line $y = x + 1$ meet at the points:

A $(-2, -1)$ and $(-1, 0)$
B $(2, 3)$ and $(1, 2)$
C $(-1, -2)$ and $(0, -1)$
D $(-2, -1)$ and $(2, 3)$

Question 152

The equation of the tangent to the
circle $x^2 + y^2 + 10x + 4y + 19 = 0$ at
the point $(-4, 1)$ is:

A $y - 3x + 7 = 0$
B $3y + x + 1 = 0$
C $y - 3x + 1 = 0$
D $3y + x - 7 = 0$

Unit 3

43

Section 1 Vectors

> **In this section you need to remember:**

1 A *vector* is a quantity which has size (magnitude) and direction. Common examples of vector quantities are velocity, acceleration, and force.

2 A vector can be shown as a line on a diagram and named by letters such as $\vec{AB}$, $\vec{CD}$, or by symbols in bold type such as **u**, **v**.

In this diagram   vector $\vec{AB}$ has components $\begin{pmatrix} 2 \\ 4 \end{pmatrix}$,

vector $\vec{CD}$ has components $\begin{pmatrix} 2 \\ -4 \end{pmatrix}$

vector **u** has components $\begin{pmatrix} -2 \\ 3 \end{pmatrix}$

vector **v** has components $\begin{pmatrix} -2 \\ -3 \end{pmatrix}$.

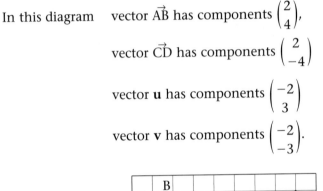

3 The *magnitude* of a vector is found by applying Pythagoras' Theorem to the components. For example, the magnitude of vector $\vec{AB}$ above is $\sqrt{2^2 + 4^2} = \sqrt{20}$.

This is written as $|\vec{AB}| = \sqrt{20}$. It is often called the *modulus* of $\vec{AB}$.

4 Vectors can be added, subtracted or multiplied by a scalar (a number).

If $\mathbf{u} = \begin{pmatrix} 6 \\ 4 \end{pmatrix}$ and $\mathbf{v} = \begin{pmatrix} -3 \\ 2 \end{pmatrix}$, then $\mathbf{u} + \mathbf{v} = \begin{pmatrix} 3 \\ 6 \end{pmatrix}$. Also, $\mathbf{u} - \mathbf{v} = \begin{pmatrix} 9 \\ 2 \end{pmatrix}$,

and $5\mathbf{u} = \begin{pmatrix} 30 \\ 20 \end{pmatrix}$.

Remember that you add or subtract **both** components, or multiply **both** components by the number.

5 Vector $\vec{BA}$ is the negative of vector $\vec{AB}$. The zero vector is $\begin{pmatrix} 0 \\ 0 \end{pmatrix}$.

6 If $\mathbf{u} = k\mathbf{v}$, where k is a constant, then vector $\mathbf{u}$ is parallel to vector $\mathbf{v}$.

7 Vectors in three dimensions follow the *same rules* as vectors in two dimensions. They have components x, y and z measured in the directions of the axes shown.

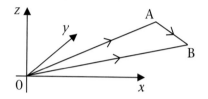

$\vec{OA}$ is called the *position vector* of a point A.

$\vec{OB}$ is the *position vector* of a point B.

Remember that vectors can be added by putting them in nose to tail arrangement so $\vec{OA} + \vec{AB} = \vec{OB}$ and so the vector $\vec{AB}$ is equal to $\vec{OB} - \vec{OA}$.

8 Points A, B, and C are collinear (lie on the same straight line) if $\vec{AB} = k\vec{BC}$.

For example, if A is the point $(2, -3, 5)$, B is $(9, -6, 9)$, and C is $(23, -12, 17)$, then

$$\vec{AB} = \vec{OB} - \vec{OA} = \begin{pmatrix} 9 \\ -6 \\ 9 \end{pmatrix} - \begin{pmatrix} 2 \\ -3 \\ 5 \end{pmatrix} = \begin{pmatrix} 7 \\ -3 \\ 4 \end{pmatrix},$$

$$\vec{BC} = \vec{OC} - \vec{OB} = \begin{pmatrix} 23 \\ -12 \\ 17 \end{pmatrix} - \begin{pmatrix} 9 \\ -6 \\ 9 \end{pmatrix} = \begin{pmatrix} 14 \\ -6 \\ 8 \end{pmatrix}.$$

It can be seen that $\vec{BC} = 2\,\vec{AB}$ so points A, B, and C are collinear.

In this example above, point B divides the line AC in the ratio $1 : 2$.

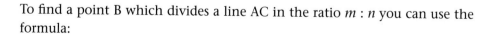

To find a point B which divides a line AC in the ratio $m : n$ you can use the formula:

$$\vec{OB} = \frac{n}{m + n}\,\vec{OA} + \frac{m}{m + n}\,\vec{OC}.$$

continued ➤

9 The unit vectors **i**, **j**, and **k** are vectors with magnitude 1 in the directions of the x, y, and z axes respectively. That is,

$$\mathbf{i} = \begin{pmatrix} 1 \\ 0 \\ 0 \end{pmatrix} \quad \mathbf{j} = \begin{pmatrix} 0 \\ 1 \\ 0 \end{pmatrix} \quad \mathbf{k} = \begin{pmatrix} 0 \\ 0 \\ 1 \end{pmatrix}.$$

10 For two vectors **a** and **b**, the *scalar product* $\mathbf{a} \cdot \mathbf{b} = |\mathbf{a}| |\mathbf{b}| \cos \theta$ where θ is the angle between **a** and **b** and $0° \le \theta \le 180°$.

Remember that if the scalar product $\mathbf{a} \cdot \mathbf{b} = 0$, the vectors **a** and **b** are perpendicular.

The angle between vectors **a** and **b** may be found using $\cos \theta = \dfrac{\mathbf{a} \cdot \mathbf{b}}{|\mathbf{a}| |\mathbf{b}|}$.

11 If $\mathbf{a} = \begin{pmatrix} a_1 \\ a_2 \\ a_3 \end{pmatrix}$ and $\mathbf{b} = \begin{pmatrix} b_1 \\ b_2 \\ b_3 \end{pmatrix}$, the scalar product $\mathbf{a} \cdot \mathbf{b} = a_1 b_1 + a_2 b_2 + a_3 b_3$.

(This is known as the *component form* of the scalar product.)

12 Note too: $\mathbf{a} \cdot \mathbf{b} = \mathbf{b} \cdot \mathbf{a}$ and $\mathbf{a} \cdot (\mathbf{b} + \mathbf{c}) = \mathbf{a} \cdot \mathbf{b} + \mathbf{a} \cdot \mathbf{c}$.

Revision questions for Unit 3 Section 1:

Question 153
If **u** and **v** are vectors such that

$\mathbf{u} + \mathbf{v} = \begin{pmatrix} 6 \\ -3 \end{pmatrix}$ and $\mathbf{u} = \begin{pmatrix} 4 \\ -1 \end{pmatrix}$, then

v is equal to:

A $\begin{pmatrix} 2 \\ -2 \end{pmatrix}$

B $\begin{pmatrix} -2 \\ 2 \end{pmatrix}$

C $\begin{pmatrix} -7 \\ -7 \end{pmatrix}$

D $\begin{pmatrix} 2 \\ 4 \end{pmatrix}$

Question 154
If **p** and **q** are vectors such that

$\mathbf{p} = \begin{pmatrix} 5 \\ -1 \end{pmatrix}$ and $\mathbf{q} = \begin{pmatrix} -6 \\ -3 \end{pmatrix}$, then $\mathbf{p} + \mathbf{q}$ is equal to:

A $\begin{pmatrix} -1 \\ -2 \end{pmatrix}$

B $\begin{pmatrix} 11 \\ 2 \end{pmatrix}$

C $\begin{pmatrix} 1 \\ 2 \end{pmatrix}$

D $\begin{pmatrix} -1 \\ -4 \end{pmatrix}$

Revision questions for Unit 3 Section 1: Continued

Question 155

If $p = \begin{pmatrix} 2 \\ -3 \end{pmatrix}$ and $q = \begin{pmatrix} 1 \\ 6 \end{pmatrix}$, then the

magnitude of $p + q$ is equal to:

A $\sqrt{90}$
B $\sqrt{18}$
C $\sqrt{12}$
D $\sqrt{6}$

Question 156

The vectors u and v are represented in the diagram. The vector sum $u + v$ has components:

A $\begin{pmatrix} 3 \\ -1 \end{pmatrix}$

B $\begin{pmatrix} -1 \\ -9 \end{pmatrix}$

C $\begin{pmatrix} 4 \\ 0 \end{pmatrix}$

D $\begin{pmatrix} 3 \\ -9 \end{pmatrix}$

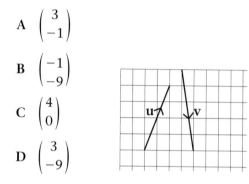

Question 157

If $u = \begin{pmatrix} -6 \\ 2 \end{pmatrix}$ and $v = \begin{pmatrix} 4 \\ -5 \end{pmatrix}$, then

$u - v$ has components:

A $\begin{pmatrix} -10 \\ -3 \end{pmatrix}$

B $\begin{pmatrix} -2 \\ -3 \end{pmatrix}$

C $\begin{pmatrix} -10 \\ 7 \end{pmatrix}$

D $\begin{pmatrix} 10 \\ -7 \end{pmatrix}$

Question 158

If $p = \begin{pmatrix} -7 \\ 5 \end{pmatrix}$ and $q = \begin{pmatrix} 8 \\ -7 \end{pmatrix}$ then $q - p$

has components:

A $\begin{pmatrix} -15 \\ 2 \end{pmatrix}$

B $\begin{pmatrix} 15 \\ 2 \end{pmatrix}$

C $\begin{pmatrix} 1 \\ -2 \end{pmatrix}$

D $\begin{pmatrix} 15 \\ -12 \end{pmatrix}$

continued ➤

Revision questions for Unit 3 Section 1: Continued

Question 159

If vector $\mathbf{p} = \begin{pmatrix} 2 \\ 1 \end{pmatrix}$ and vector $\mathbf{q} = \begin{pmatrix} 3 \\ 0 \end{pmatrix}$,

then $3\mathbf{p} - 2\mathbf{q}$ has components:

A $\begin{pmatrix} 0 \\ 3 \end{pmatrix}$

B $\begin{pmatrix} 3 \\ -2 \end{pmatrix}$

C $\begin{pmatrix} 5 \\ -2 \end{pmatrix}$

D $\begin{pmatrix} 0 \\ 1 \end{pmatrix}$

Question 160

If vector $\mathbf{a} = \begin{pmatrix} -2 \\ 4 \end{pmatrix}$ and vector $\mathbf{b} = \begin{pmatrix} -3 \\ 1 \end{pmatrix}$,

the magnitude of the vector $\mathbf{a} + 2\mathbf{b}$ is:

A $\sqrt{52}$

B 2

C 10

D $\sqrt{50}$

Question 161

In the diagram vector $\overrightarrow{PQ}$ is equal to:

A $\overrightarrow{OP} + \overrightarrow{OQ}$
B $\overrightarrow{OP} - \overrightarrow{OQ}$
C $\overrightarrow{OQ} - \overrightarrow{OP}$
D $\overrightarrow{OQ} + \overrightarrow{OP}$

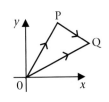

Question 162

If E is the point $(-6, 1)$ and F is the point $(2, 5)$ then vector $\overrightarrow{EF}$ has components:

A $\begin{pmatrix} 8 \\ 4 \end{pmatrix}$

B $\begin{pmatrix} -4 \\ 6 \end{pmatrix}$

C $\begin{pmatrix} -8 \\ -4 \end{pmatrix}$

D $\begin{pmatrix} 4 \\ 8 \end{pmatrix}$

Revision questions for Unit 3 Section 1: Continued

Question 163
In the diagram vector $\vec{PQ}$ has components:

A $\begin{pmatrix} -6 \\ -7 \end{pmatrix}$

B $\begin{pmatrix} 4 \\ 7 \end{pmatrix}$

C $\begin{pmatrix} 6 \\ -1 \end{pmatrix}$

D $\begin{pmatrix} -4 \\ -7 \end{pmatrix}$

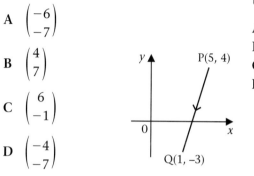

Question 164
If the points T(2, −1), U(−1, 1), and V(k, 3) lie on the same straight line then k is equal to:

A −3
B −4
C 3
D 4

Question 165
If A is the point (2, −1) and B is the point (−4, 5), then P, the point which divides the line AB in the ratio 2:1, has coordinates:

A (0, 1)
B (−1, 2)
C (−2, 3)
D (3, −2)

Question 166
If P is the point (−2, 3) and Q is the point (7, −3), the point R which divides PQ in the ratio 2:1 has coordinates:

A (4, −1)
B (1, 1)
C $(2\frac{1}{2}, 0)$
D (6, −4)

Question 167
In the diagram PQ represents vector **v** and QR represents vector **w**. If S divides PR in the ratio 5:3 then SR represents the vector:

A $\frac{5}{8}(\mathbf{w} - \mathbf{v})$

B $\frac{3}{8}(\mathbf{w} - \mathbf{v})$

C $\frac{3}{8}(\mathbf{v} + \mathbf{w})$

D $\frac{5}{8}(\mathbf{v} + \mathbf{w})$

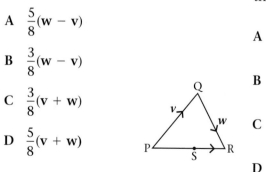

Question 168
If vector $\vec{AB} = \begin{pmatrix} 20 \\ -5 \end{pmatrix}$, and the point P divides the line AB in the ratio 4:1, then vector $\vec{AP}$ has components:

A $\begin{pmatrix} 4 \\ -1 \end{pmatrix}$

B $\begin{pmatrix} 16 \\ -4 \end{pmatrix}$

C $\begin{pmatrix} 5 \\ -1\frac{1}{4} \end{pmatrix}$

D $\begin{pmatrix} 16 \\ 4 \end{pmatrix}$

continued ➤

Revision questions for Unit 3 Section 1: Continued

Question 169

In terms of the unit vectors **i**, **j**, and **k**,

the vector $\overrightarrow{PQ} = \begin{pmatrix} 2 \\ -5 \\ 6 \end{pmatrix}$ can be written as:

A $6\mathbf{i} - 5\mathbf{j} - 2\mathbf{k}$

B $2, -5, 6$

C $\dfrac{2}{\sqrt{65}}\mathbf{i} - \dfrac{5}{\sqrt{65}}\mathbf{j} + \dfrac{6}{\sqrt{65}}\mathbf{k}$

D $2\mathbf{i} - 5\mathbf{j} + 6\mathbf{k}$

Question 170

If $\mathbf{t} = 2\mathbf{i} - 3\mathbf{j} + 4\mathbf{k}$, and $\mathbf{u} = \mathbf{i} + 2\mathbf{j} - \mathbf{k}$, then in component form $2\mathbf{t} - \mathbf{u}$ is:

A $\begin{pmatrix} 3 \\ -8 \\ 9 \end{pmatrix}$

B $\begin{pmatrix} 3 \\ -4 \\ 1 \end{pmatrix}$

C $\begin{pmatrix} 3 \\ -5 \\ 2 \end{pmatrix}$

D $\begin{pmatrix} -3 \\ -8 \\ 1 \end{pmatrix}$

Question 171

If $\mathbf{m} = \begin{pmatrix} 2 \\ -6 \\ 1 \end{pmatrix}$ and $\mathbf{n} = \begin{pmatrix} 3 \\ 1 \\ 0 \end{pmatrix}$ then

$2\mathbf{m} - \mathbf{n}$ is:

A $\begin{pmatrix} 1 \\ -13 \\ 2 \end{pmatrix}$

B $\begin{pmatrix} 2 \\ -14 \\ 2 \end{pmatrix}$

C $\begin{pmatrix} 1 \\ -7 \\ 1 \end{pmatrix}$

D $\begin{pmatrix} 1 \\ -11 \\ 2 \end{pmatrix}$

Question 172

Vector $\mathbf{p} = \begin{pmatrix} -4 \\ 2 \\ 1 \end{pmatrix}$ and vector $\mathbf{q} = \begin{pmatrix} h \\ 1 \\ -5 \end{pmatrix}$.

If the magnitude of $\mathbf{p} + \mathbf{q} = 5$ then h is:

A   1

B   10

C   4

D   6

Revision questions for Unit 3 Section 1: Continued

Question 173

If K is the point $(-1, 3, 0)$ and L is the point $(4, 3, 2)$, then vector $\vec{KL}$ has components:

A $\begin{pmatrix} 3 \\ 6 \\ 2 \end{pmatrix}$

B $\begin{pmatrix} -3 \\ 0 \\ 2 \end{pmatrix}$

C $\begin{pmatrix} 5 \\ 0 \\ 2 \end{pmatrix}$

D $\begin{pmatrix} -5 \\ 0 \\ -2 \end{pmatrix}$

Question 174

If G is the point $(3, -2, 4)$ and H is the point $(-7, 4, -1)$, then vector $\vec{GH}$ has components:

A $\begin{pmatrix} -10 \\ 6 \\ -5 \end{pmatrix}$

B $\begin{pmatrix} -4 \\ 2 \\ 3 \end{pmatrix}$

C $\begin{pmatrix} -10 \\ 6 \\ 5 \end{pmatrix}$

D $\begin{pmatrix} 10 \\ -6 \\ 5 \end{pmatrix}$

Question 175

If M is the point $(-3, 3, -2)$, N is the point $(7, -2, 18)$, and if MT:TN = 2:3, then the point T has coordinates:

A $(2, -1, 8)$
B $(-7, 5, -10)$
C $(3, 0, 10)$
D $(1, 1, 6)$

Question 176

If P is the point $(0, -6, 6)$, Q is the point $(12, 18, -6)$, and if T divides PQ in the ratio 3:1, then the point T has coordinates:

A $(4, 0, 3)$
B $(9, 12, -3)$
C $(9, 12, 15)$
D $(-9, -24, 15)$

Question 177

If TUVW is a parallelogram with vertices T $(2, 3, -1)$, U $(7, 6, -4)$ and V $(7, 4, 0)$ then W has coordinates:

A $(12, 7, -3)$
B $(5, 3, -3)$
C $(2, 1, 3)$
D $(5, -3, 3)$

Question 178

Three points P $(2, 7, 11)$, Q $(3, 5, 6)$ and R $(8, m, n)$ are collinear if m and n have the values:

A $m = -5; n = -19$
B $m = 5; n = 6$
C $m = 13; n = 17$
D $m = -15; n = -31$

continued ➤

Revision questions for Unit 3 Section 1: Continued

Question 179

If vectors **p** and **q** have magnitudes $|\mathbf{p}| = 6$ and $|\mathbf{q}| = 3$, and the angle between them is 60°, the scalar product **p . q** is equal to:

A   18

B $9\sqrt{3}$

C   9

D $4\frac{1}{2}$

Question 180

If vectors **t** and **v** have magnitudes $|\mathbf{t}| = 5$ and $|\mathbf{v}| = 8$, and the angle between them is 150°, the scalar product **t . v** is equal to:

A -20

B $-20\sqrt{3}$

C $20\sqrt{3}$

D   40

Question 181

If **m** is the vector $\begin{pmatrix} 3 \\ -2 \\ 1 \end{pmatrix}$ and **n** is the vector $\begin{pmatrix} 4 \\ -5 \\ -2 \end{pmatrix}$ then the scalar product

m . n is equal to:

A   0

B -1

C $\sqrt{(14 \times 45)}$

D   20

Question 182

If $\vec{PQ}$ has components $\begin{pmatrix} 2 \\ 2 \\ -1 \end{pmatrix}$ and $\vec{PS}$

has components $\begin{pmatrix} -5 \\ 3 \\ -4 \end{pmatrix}$,

then the

scalar product $\vec{PQ}.\vec{PS}$ is:

A -8

B   0

C -2

D $\sqrt{450}$

Question 183

If $\vec{PQ}$ is the vector $\begin{pmatrix} 2 \\ 0 \\ -2 \end{pmatrix}$ and $\vec{PR}$ is

$\begin{pmatrix} 0 \\ 0 \\ -2 \end{pmatrix}$ then the cosine of angle RPQ is:

A $\dfrac{1}{\sqrt{2}}$

B $\dfrac{\sqrt{3}}{2}$

C $\sqrt{3}$

D $\sqrt{2}$

Question 184

If $\vec{GH} = \begin{pmatrix} 2 \\ -3 \\ 5 \end{pmatrix}$ and $\vec{KL} = \begin{pmatrix} 1 \\ 4 \\ 2 \end{pmatrix}$ then the

size of angle HGL is:

A   0°

B   180°

C   90°

D   unknown without further information.

Revision questions for Unit 3 Section 1: Continued

Question 185

Vectors $\begin{pmatrix} 7 \\ 3 \\ -2 \end{pmatrix}$ and $\begin{pmatrix} 2 \\ p \\ 1 \end{pmatrix}$ are

perpendicular if p has the value:

A $-3\dfrac{2}{3}$

B 4

C -4

D -12

Question 186

The points P, Q, and R are shown in the diagram. The cosine of the angle θ between PQ and PR has the value:

A $\dfrac{5}{3\sqrt{6}}$

B $\dfrac{5}{\sqrt{3}}$

C $\dfrac{\sqrt{5}}{3\sqrt{6}}$

D $\dfrac{6}{3\sqrt{6}}$

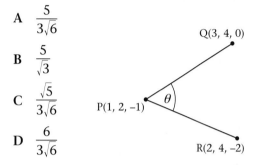

Q(3, 4, 0)

P(1, 2, –1) θ

R(2, 4, –2)

Section 2 Further calculus

> ### In this section you need to remember:
>
> **1** To differentiate $\sin x$ and $\cos x$ use the following rules:
>
> if $y = \sin x$ $\dfrac{dy}{dx} = \cos x,$
>
> if $y = \cos x$ $\dfrac{dy}{dx} = -\sin x.$
>
> **2** To integrate $\sin x$ and $\cos x$ use the following rules:
>
> if $y = \sin x$ $\displaystyle\int y\,dx = \int \sin x\,dx = -\cos x + C$
>
> if $y = \cos x$ $\displaystyle\int y\,dx = \int \cos x\,dx = \sin x + C.$
>
> **3** If $f(x) = (x + 5)^n$ $f'(x) = n(x + 5)^{n-1}$
>
> If $f(x) = (3x + 5)^n$ $f'(x) = 3n(3x + 5)^{n-1}.$
>
> **4** If $y = (x^2 + 4x)^{\frac{1}{2}}$ first put $u = x^2 + 4x$ so that $y = u^{\frac{1}{2}}.$
>
> Now $\dfrac{dy}{dx} = \dfrac{dy}{du} \times \dfrac{du}{dx}$ (The chain rule).
>
> $\dfrac{dy}{du} = \dfrac{1}{2}\, u^{-\frac{1}{2}};\ \dfrac{du}{dx} = 2x + 4$
>
> so $\dfrac{dy}{dx} = \dfrac{1}{2}\, u^{-\frac{1}{2}} \times (2x + 4)$
>
> $\phantom{so \dfrac{dy}{dx}} = \dfrac{1}{2}(x^2 + 4x)^{-\frac{1}{2}}(2x + 4)$
>
> $\phantom{so \dfrac{dy}{dx}} = \dfrac{2x + 4}{2(x^2 + 4x)^{\frac{1}{2}}}$
>
> **5** $\displaystyle\int (3x + 5)^3\,dx$ and other similar integrals can be found because of the chain rule.
>
> $\displaystyle\int (3x + 5)^3\,dx = (3x + 5)^4$ divided by the new power and by the coefficient of x.
>
> That is, $\dfrac{1}{12}\,(3x + 5)^4.$

6 $\int \sin(ax + b)\, dx = -\dfrac{1}{a} \cos(ax + b) + C.$

7 $\int \cos(ax + b)\, dx = \dfrac{1}{a} \sin(ax + b) + C.$

Revision questions for Unit 3 Section 2:

Question 187

If $f(x) = 2 \cos x - 3 \sin x$, the derivative of $f(x)$ is:

A $2 \sin x + 3 \cos x$

B $2 \sin x - 3 \cos x$

C $-2 \sin x + 3 \cos x$

D $-2 \sin x - 3 \cos x$

Question 188

If $g(x) = 2 \sin x - 6 \cos x$, then $g'(x)$ is:

A $2 \cos x - 6 \sin x$

B $2 \cos x + 6 \sin x$

C $-2 \cos x + 6 \sin x$

D $-2 \cos x - 6 \sin x$

Question 189

The tangent to the curve $y = 3 \cos x$ at the point where $x = \dfrac{1}{3}\pi$ has gradient:

A $-\dfrac{3}{2}\sqrt{3}$

B $-\dfrac{3}{2}$

C $\dfrac{3}{2}\sqrt{3}$

D $\dfrac{3}{2}$

Question 190

The tangent to the curve $y = 5 \sin x$ at the point where $x = \dfrac{5}{6}\pi$ has gradient:

A $\dfrac{-5\sqrt{3}}{2}$

B $\dfrac{5\sqrt{3}}{2}$

C $\dfrac{-5}{2}$

D $\dfrac{5}{2}$

Question 191

The integral $\int (4 \cos x + 9 \sin x)\, dx$ is equal to:

A $4 \sin x - 9 \cos x + C$

B $-4 \sin x + 9 \cos x + C$

C $-4 \sin x - 9 \cos x + C$

D $4 \sin x + 9 \cos x + C$

Question 192

The integral $\int (6 \sin \theta - 3 \cos \theta)\, d\theta$ is equal to:

A $6 \cos \theta - 3 \sin \theta + C$

B $-6 \cos \theta + 3 \sin \theta + C$

C $6 \cos \theta + 3 \sin \theta + C$

D $-6 \cos \theta - 3 \sin \theta + C$

continued ➤

Revision questions for Unit 3 Section 2: Continued

Question 193

The integral $\int_0^{2\pi} \sin x \, dx$ is equal to:

A $\quad -\dfrac{1}{2}$

B $\quad 0$

C $\quad \dfrac{\sqrt{3}}{2}$

D $\quad -\dfrac{\sqrt{3}}{2}$

Question 194

The integral $\int_0^{\pi} \sin x \, dx$ is equal to:

A $\quad 0$

B $\quad 1$

C $\quad -2$

D $\quad 2$

Question 195

The integral $\int_0^{\frac{\pi}{4}} (5 \cos x - \sin x) \, dx$
is equal to:

A $\quad 3\sqrt{2} - 1$
B $\quad 3\sqrt{2} + 1$
C $\quad 6\sqrt{2} - 1$
D $\quad 6\sqrt{2} + 1$

Question 196

The integral $\int_{\frac{\pi}{2}}^{\pi} (5 \sin x + x^2) \, dx$
is equal to:

A $\quad 5 + \dfrac{7\pi^3}{24}$

B $\quad 5 + \dfrac{9\pi^3}{24}$

C $\quad 5 + \pi$

D $\quad -5 + \dfrac{7\pi^3}{24}$

Question 197

The derivative of $(3x + 2)^5$ is equal to:

A $\quad 5(3x + 2)^4$

B $\quad \dfrac{5}{3}(3x + 2)^4$

C $\quad 15(3x + 2)^4$

D $\quad \dfrac{1}{2}(3x + 2)^6$

Question 198

If $f(x) = \sqrt{(2x + 3)}$ then $f'(x)$ is equal to:

A $\quad \dfrac{1}{2}\sqrt{(2x + 3)^3}$

B $\quad \dfrac{1}{\sqrt{(2x + 3)}}$

C $\quad \dfrac{1}{3}\sqrt{(2x + 3)^{\frac{3}{2}}}$

D $\quad \dfrac{1}{2\sqrt{(2x + 3)}}$

Revision questions for Unit 3 Section 2: Continued

Question 199

If $g(x) = \cos(4x + 3)$, then $g'(x)$ is equal to:

A $-\sin(4x + 3)$

B $4\sin(4x + 3)$

C $-\dfrac{1}{4}\sin(4x + 3)$

D $-4\sin(4x + 3)$

Question 200

If $y = \cos^3\theta$, then $\dfrac{dy}{d\theta}$ is equal to:

A $3\cos^2\theta \sin\theta$

B $3\sin^2\theta$

C $-3\cos^2\theta$

D $-3\cos^2\theta \sin\theta$

Question 201

If $y = \dfrac{1}{\cos^2 x}$, then $\dfrac{dy}{dx}$ is equal to:

A $\dfrac{2\sin x}{\cos^3 x}$

B $\dfrac{1}{2\cos x}$

C $\dfrac{-2\sin x}{\cos^3 x}$

D $\dfrac{1}{2\sin x \cos x}$

Question 202

If $y = \cos\sqrt{x}$ then $\dfrac{dy}{dx}$ is equal to:

A $-\dfrac{1}{2}\sin\sqrt{x}$

B $\dfrac{-\sin\sqrt{x}}{2\sqrt{x}}$

C $-\dfrac{2}{3}\sin\sqrt{x}$

D $\dfrac{\sin\sqrt{x}}{2\sqrt{x}}$

Question 203

The integral $\displaystyle\int (1 - 2t)^3 \, dt$ is equal to:

A $-\dfrac{1}{8}(1 - 2t)^4 + C$

B $-\dfrac{1}{2}(1 - 2t)^4 + C$

C $-6(1 - 2t)^2 + C$

D $-\dfrac{1}{4}(1 - 2t)^4 + C$

Question 204

The integral of $\dfrac{1}{(2x + 1)^2}$ with respect to x is:

A $\dfrac{1}{2(2x + 1)} + C$

B $\dfrac{-1}{2(2x + 1)} + C$

C $\dfrac{2}{(2x + 1)} + C$

D $\dfrac{-2}{(2x + 1)} + C$

continued ➤

Revision questions for Unit 3 Section 2: Continued

Question 205

The integral $\int_0^{\frac{\pi}{6}} \cos 3x \, dx$ is equal to:

A $-\dfrac{1}{3}$

B $\dfrac{1}{6}$

C $\dfrac{1}{3}$

D $-\dfrac{1}{6}$

Question 206

The integral $\int_0^{\frac{\pi}{2}} (\cos 2x - \sin 2x) \, dx$ is equal to:

A 0

B $-\dfrac{1}{2}$

C $\dfrac{1}{2}$

D -1

Section 3 Exponential and logarithmic functions

In this section you need to remember:

1 The logarithm of a number to a given base is the power to which the base has to be raised to give the number.

If $\log_a y = x$ then $y = a^x$.

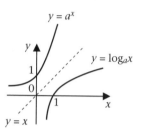

The graphs of these functions are reflections of each other in the line $y = x$. Each function is the *inverse function* of the other.

2 The rules for working with logarithms are:

$$\log_a xy = \log_a x + \log_a y$$

$$\log_a \frac{x}{y} = \log_a x - \log_a y$$

$$\log_a x^n = n\log_a x.$$

3 The number e which has the value $2 \cdot 718\ldots$ is used as the base of logarithms in many applications. If $y = e^x$, then $\log_e y = x$.

4 If in an experiment the graph of results showing $\log y$ against $\log x$ is a straight line, its equation can be written as $\log y = m\log x + c$ or $\log y = m\log x + \log k$ (where $c = \log k$).

This equation can be written as $\log y = \log x^m + \log k$

or $\log y = \log kx^m$

so $y = kx^m$.

continued ➤

5 If in an experiment the results show log y against x is a straight line, then its equation can be written as $\log y = mx + c$ or $\log y = x\log b + \log a$ (where log b is then the gradient of the line and log $a = c$)

so $\log y = \log ab^x$ and therefore $y = ab^x$.

6 For any value of the base a, $\log_a a = 1$, and $\log_a 1 = 0$.

Revision questions for Unit 3 Section 3:

Question 207
The relationship $y = a^x$ written in logarithmic form is:

A $\log_x a = y$
B $\log_a y = x$
C $\log_x y = a$
D $\log_a x = y$

Question 208
If $k = \log_r m$ then:

A $m^k = r$
B $r^m = k$
C $m^r = k$
D $r^k = m$

Question 209
If $\log_b y = \log_b 3 - 2\log_b x$, then y expressed in terms of x is:

A $y = \dfrac{3}{x^2}$

B $y = 3 - x^2$

C $y = \dfrac{3}{2x}$

D $y = 3 - 2x$

Question 210
If $\log_p t = 3\log_p v + \log_p 5$, then t expressed in terms of v is:

A $t = 3v + 5$
B $t = v^3 + 5$
C $t = 5v^3$
D $t = 15v$

Question 211
The equation
$\log_a x + \log_a(2x + 9) = \log_a 5 \ (x > 0)$
has solution:

A $x = \dfrac{1}{2}$

B $x = -\dfrac{4}{3}$

C $x = \dfrac{5}{2}$ or $x = 1$

D $x = \dfrac{1}{2}$ or $x = -5$

Question 212
The equation
$\log_b(x - 1) + \log_b(x + 5) = \log_b 7 \ (x > 0)$
has solution:

A $x = \dfrac{3}{2}$

B $x = 2$
C $x = 7$ or $x = 3$
D $x = 8$ or $x = 2$

Revision questions for Unit 3 Section 3: Continued

Question 213

Solving the equation

$\log_b(x - 6) + \log_b(x - 4) = \log_b x (x > 0)$, gives:

A $x = 10$
B $x = 6$ or $x = 4$
C $x = 12$ or $x = 2$
D $x = 3$ or $x = 8$

Question 214

Solving the equation

$\log_{10}x + \log_{10}(2x - 1) = 10 (x > 0)$ gives:

A $x = 2\frac{1}{2}$ or $x = -2$

B $x = 1$

C $x = \frac{1}{2}$

D $x = -\frac{1}{2}$ or $x = 1$

Question 215

If after an experiment the results can be plotted as shown in the diagram, the formula connecting y and x will be of the form:

A $y = nx + k$
B $y = kn^x$
C $y = nx^k$
D $y = kx^n$

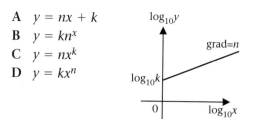

Question 216

If after an experiment the results can be plotted as shown in the diagram, the formula connecting y and x will be of the form:

A $y = ax + b$
B $y = a + b^x$
C $y = ab^x$
D $y = a + bx$

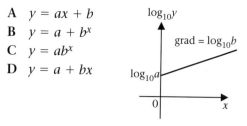

Question 217

If the full line in the diagram shows $y = \log_e x$, then the dotted line is most likely to show:

A $y = \log_e(-x)$
B $y = e^x$
C $y = -\log_e x$

D $y = \dfrac{1}{\log_e x}$

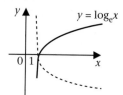

Question 218

If the full line in the diagram shows $y = \log_e x$, then the dotted line is most likely to show:

A $y = \log_e(x + 2)$
B $y = -2 + \log_e x$
C $y = \log_e(x - 2)$
D $y = 2 - \log_e x$

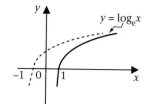

Section 4 The Wave Function

In this section you need to remember:

1 An expression of the form $a \sin x + b \cos x$ can be rewritten to find its maximum or minimum value, or to help solve equations. One of the following can be used.

- $a \sin x + b \cos x = k \sin (x + \alpha) = k \sin x \cos \alpha + k \cos x \sin \alpha$

  so $k \cos \alpha = a$, and $k \sin \alpha = b$

- $a \sin x - b \cos x = k \sin (x - \alpha) = k \sin x \cos \alpha - k \cos x \sin \alpha$

  so $k \cos \alpha = a$, and $k \sin \alpha = b$

- $b \cos x + a \sin x = k \cos (x + \alpha) = k \cos x \cos \alpha - k \sin x \sin \alpha$

  so $k \cos \alpha = b$, and $k \sin \alpha = -a$

- $b \cos x - a \sin x = k \cos (x - \alpha) = k \cos x \cos \alpha + k \sin x \sin \alpha$

  so $k \cos \alpha = b$, and $k \sin \alpha = -a$.

The value of k can be found from $k^2 (\sin^2\alpha + \cos^2\alpha) = a^2 + b^2$,

and $\tan \alpha$ can be found from $\dfrac{\sin \alpha}{\cos \alpha}$, as in the example in 2 below.

2 The maximum value of an expression $k \sin (x + \alpha)$ must be k since $\sin (x + \alpha)$ cannot be more than 1.

The minimum value of $k \sin (x + \alpha)$ must be $-k$ since $\sin (x + \alpha)$ cannot be less than -1. (The same can be said of the other expressions in 1 above.)

For example, to express $\sin x° - \cos x°$ in the form $k \cos(x - \alpha)°$ and find its maximum value, we first write

$$k \cos (x - \alpha)° = k \cos x° \cos \alpha° + k \sin x° \sin \alpha°$$

Then, $k \cos \alpha° = -1$ and $k \sin \alpha° = 1$.

So $\tan \alpha° = -1$ and $\alpha°$ must be in the second quadrant, because $\sin \alpha$ is positive and $\cos \alpha$ is negative.

Hence $\alpha = 135$.

Now $k = \sqrt{(-1)^2 + 1^2} = \sqrt{2}$ (from $k^2\sin^2\alpha + k^2\cos^2\alpha = k^2(\sin^2\alpha + \cos^2\alpha) = k^2$)

so $\sin x° - \cos x° = \sqrt{2}\cos(x - 135)°$, which has a maximum value of $\sqrt{2}$.

This value occurs when $(x - 135)°$ equals zero, therefore when $x = 135$.

3 The same approach can be used for solving equations such as
$\sin x° - \cos x° = 1$ $(0 \leq x \leq 360)$.

This leads to $\sqrt{2}\cos(x - 135)° = 1$

$$\Rightarrow \cos(x - 135)° = \frac{1}{\sqrt{2}}.$$

If $0 \leq x \leq 360$, this means that $-135° \leq (x - 135)° \leq 225°$,

and so $(x - 135)° = -45°$ or $45°$ (think about the graph of $\cos x$).

Therefore $x = 90$ or 180.

Revision questions for Unit 3 Section 4:

Question 219

When $5\sin x° + 3\cos x°$ is written in the form $k\sin(x - \alpha)°$ where $0 \leq \alpha \leq 360$, $\tan \alpha°$ is equal to:

A $\dfrac{5}{3}$

B $-\dfrac{5}{3}$

C $-\dfrac{3}{5}$

D $\dfrac{3}{5}$

Question 220

When $6\sin x° - 7\cos x°$ is written in the form $k\sin(x + \alpha)°$ where $0 \leq \alpha \leq 360$, $\tan \alpha°$ is equal to:

A $\dfrac{6}{7}$

B $-\dfrac{7}{6}$

C $\dfrac{7}{6}$

D $-\dfrac{6}{7}$

continued ➤

Revision questions for Unit 3 Section 4: Continued

Question 221

When $8 \cos x° + 7 \sin x°$ is written in the form $k \cos (x - \alpha)°$, $0 \leq \alpha \leq 360$, $\tan \alpha°$ is equal to:

A $-\dfrac{8}{7}$

B $-\dfrac{7}{8}$

C $\dfrac{8}{7}$

D $\dfrac{7}{8}$

Question 222

When $4 \cos x° - 5 \sin x°$ is written in the form $k \cos (x + \alpha)°$, $0 \leq \alpha \leq 360$, $\tan \alpha°$ is equal to:

A $-\dfrac{5}{4}$

B $\dfrac{5}{4}$

C $\dfrac{4}{5}$

D $-\dfrac{4}{5}$

Question 223

If $k \sin \alpha° = -\sqrt{3}$ and $k \cos \alpha° = 1$, when $0 \leq \alpha \leq 360$ and $k > 0$, then k and α have the values:

A $k = 2$, $\alpha = 120$ or 300
B $k = 4$, $\alpha = 120$ or 300
C $k = 4$, $\alpha = 300$
D $k = 2$, $\alpha = 300$

Question 224

If $k \sin \alpha = 4$ and $k \cos \alpha = 4\sqrt{3}$, when $0 \leq \alpha \leq 2\pi$ and $k > 0$, then k and α have the values:

A $k = 8$, $\alpha = \dfrac{\pi}{6}$

B $k = \sqrt{28}$, $\alpha = \dfrac{\pi}{6}$

C $k = 64$, $\alpha = \dfrac{\pi}{6}$ or $\dfrac{7\pi}{6}$

D $k = 8$, $\alpha = \dfrac{\pi}{6}$ or $\dfrac{7\pi}{6}$

Question 225

The function $f(x) = \sin x° + \sqrt{3} \cos x°$ ($0 \leq x \leq 360$) has the maximum value shown for the given value of x:

A   maximum value $= \sqrt{3}$ when $x = 0$
B   maximum value $= 2$ when $x = 30$
C   maximum value $= 1$ when $x = 90$
D   maximum value $= 2$ when $x = 60$

Question 226

The function $f(\theta) = \cos \theta - \sin \theta$ ($0 \leq \theta \leq 2\pi$) has the maximum value shown for the given value of θ:

A   maximum value $= \sqrt{2}$ when $\theta = \dfrac{3\pi}{4}$

B   maximum value $= 1$ when $\theta = \dfrac{\pi}{4}$

C   maximum value $= \sqrt{2}$ when $\theta = \dfrac{7\pi}{4}$

D   maximum value $= 2$ when $\theta = \dfrac{5\pi}{4}$

Revision questions for Unit 3 Section 4: Continued

Question 227

If in solving the equation
$\sqrt{3} \sin x° + \cos x° = 1$ $(0 \le x \le 360)$,
it can be shown that $2 \sin (x + 30)° = 1$,
then x is equal to:

A   0, 120, or 360

B   0 or 120

C   30, or 90

D   30, 90, or 360

Question 228

If in solving the equation
$\cos x - \sin x = -1$ $(0 \le x \le 360)$ it
can be shown that
$\sqrt{2} \cos (x - 315)° = -1$, then:

A $x = 135$ or 225

B $x = 90$

C $x = 90$ or 180

D x has no values in the given range

Question 229

If in solving the equation
$\sqrt{3} \cos x + \sin x = \sqrt{2}$ $(0 \le x \le 2\pi)$, it

can be shown that $\cos(x - \dfrac{\pi}{6}) = \dfrac{1}{\sqrt{2}}$,

then the possible values of x are:

A $\dfrac{5\pi}{12}$ or $\dfrac{13\pi}{12}$

B $\dfrac{\pi}{12}$ or π

C $\dfrac{5\pi}{12}$ or $\dfrac{23\pi}{12}$

D $\dfrac{\pi}{4}$ or $\dfrac{7\pi}{4}$

Question 230

If in solving the equation
$\cos x - \sqrt{3} \sin x = 1$ $(0 \le x \le 2\pi)$, it

can be shown that $2 \cos(x - \dfrac{5\pi}{3}) = 1$,

then the possible values of x are:

A $\dfrac{4\pi}{3}$

B $\dfrac{11\pi}{6}$ or $\dfrac{3\pi}{2}$

C   0 or $\dfrac{\pi}{3}$

D   0, $\dfrac{4\pi}{3}$ or 2π

Formulae Sheet

(This formulae sheet is similar to that provided by the SQA for the Higher Mathematics Examination. You may find it helpful to refer to it for the six Test Sets which follow.)

Trigonometric formulae:

$\sin (A + B) = \sin A \cos B + \cos A \sin B$

$\sin (A - B) = \sin A \cos B - \cos A \sin B$

$\cos (A + B) = \cos A \cos B - \sin A \sin B$

$\cos (A - B) = \cos A \cos B + \sin A \sin B$

$\sin 2A = 2 \sin A \cos A$

$\cos 2A = \cos^2 A - \sin^2 A$

$\quad\quad = 2\cos^2 A - 1$

$\quad\quad = 1 - 2 \sin^2 A$

Circle formulae:

The equation of a circle with its centre at the origin and having radius r is $x^2 + y^2 = r^2$.

The equation of a circle with its centre at the point (a, b) and having radius r is $(x - a)^2 + (y - b)^2 = r^2$.

The general equation of a circle is $x^2 + y^2 + 2gx + 2fy + c = 0$ (g, f and c being constants).

In the general equation, the centre of the circle is the point $(-g, -f)$ and the radius has length $\sqrt{g^2 + f^2 - c}$.

Vector formulae:

For two vectors $\mathbf{a}$ and $\mathbf{b}$ the scalar product $\mathbf{a} \cdot \mathbf{b} = |\mathbf{a}||\mathbf{b}| \cos \theta$, where θ is the angle between $\mathbf{a}$ and $\mathbf{b}$ and $0° \le \theta \le 180°$.

If vector $\mathbf{a} = \begin{pmatrix} a_1 \\ a_2 \\ a_3 \end{pmatrix}$ and vector $\mathbf{b} = \begin{pmatrix} b_1 \\ b_2 \\ b_3 \end{pmatrix}$, the scalar product $\mathbf{a} \cdot \mathbf{b} = a_1b_1 + a_2b_2 + a_3b_3$.

Standard derivatives:

If $f(x) = \sin ax$, $f'(x) = a \cos ax$.

If $f(x) = \cos ax$, $f'(x) = -a \sin ax$.

Standard integrals:

If $f(x) = \sin ax$, $\int f(x) \, dx = -\frac{1}{a} \cos ax + C$.

If $f(x) = \cos ax$, $\int f(x) \, dx = \frac{1}{a} \sin ax + C$.

Test Sets 1–6

The following six Test Sets each contain twenty questions of the type and difficulty level which you can expect to meet in the objective questions part of the Higher Grade examination.

No difficulty level indicators are shown with the answers but each question is worth the equivalent of two examination marks.

Test Set 1

Question 1

The straight line through the point (2, 5) which is parallel to the line $y = 2x - 3$ has equation:

A $y + 2x = 9$
B $y - 2x = 3$
C $2y + x = 15$
D $y - 2x = 1$

Question 2

The lines with equations $3x + 2y = 7$ and $2x - 2y = 8$ meet at the point:

A $(3, -1)$
B $(3, 1)$
C $(-1, 3)$
D $(-1, 5)$

Question 3

If $f(x) = 1 - 3x^2$ and $g(x) = 4x$, then $f(g(x))$ is equal to:

A $1 - 12x^2$
B $4 - 12x^2$
C $1 - 48x^2$
D $1 - 144x^2$

Question 4

The graph of the function $y = \log_5 x + 3$ passes through the point $(1, k)$ if k has the value:

A 8
B 3
C 0
D 5

Question 5

The diagram shows the graph of a trigonometric function. Its equation is:

A $y = \sin 2x + 1$
B $y = 2 \sin x + 1$
C $y = \cos 2x + 1$
D $y = \sin(2x + 1)$

Question 6

The maximum value of the function $f(x) = 2\sin 2x - 1$ is:

A 3
B 1
C -3
D 0

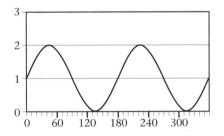

Test Set 1 continued

Question 7

In the sequence defined by $t_n = 2{\cdot}5t_{n-1} + 60$, as n tends to infinity t_n tends to:

A -40

B 40

C infinity

D 24

Question 8

The function $g(x) = x^3 - 27$ is a decreasing function for:

A $x < 3$

B all values of x

C all values of x except $x = 0$

D no values of x

Question 9

One of the factors of $x^3 - 3x^2 - 6x + 8$ is $(x + 2)$. The other factors are:

A $(x - 1)$ and $(x - 4)$

B $(x + 4)$ and $(x + 1)$

C $(x - 1)$ and $(x + 4)$

D $(x - 4)$ and $(x + 1)$

Question 10

The parabola with equation $y = (x - 5)(x + 3)$ has an axis of symmetry on the line with equation:

A $x = -2$

B $x = 1$

C $x = 2$

D $x = 4$

Question 11

The equation $2x^2 + 3x - 35 = 0$ has roots:

A $x = 2\dfrac{1}{2}$ and $x = 7$

B $x = 3\dfrac{1}{2}$ and $x = -5$

C $x = -3\dfrac{1}{2}$ and $x = 5$

D $x = 7$ and $x = -5$

Question 12

The integral $\int \left(2x - \dfrac{1}{\sqrt{x}}\right)dx$ is equal to:

A $2 + \dfrac{1}{2}\sqrt{x^3} + C$

B $x^2 - 2\sqrt{x} + C$

C $4x^2 - 4\sqrt{x} + C$

D $\dfrac{1}{2}x^2 - 2\sqrt{x} + C$

continued ➤

Test Set 1 continued

Question 13
The integral $\int_1^8 t^{\frac{1}{3}}\,dt$ has the value:

A $12\frac{3}{4}$

B $4\frac{1}{2}$

C $11\frac{1}{4}$

D 12

Question 14
In the diagram the shaded area (in units^2) is equal to:

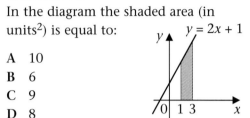

A 10

B 6

C 9

D 8

Question 15
The expression
$\sin70°\cos10° - \cos10°\sin70°$ is equal to:

A $-\dfrac{1}{2}$

B $\dfrac{\sqrt{3}}{2}$

C $\dfrac{1}{2}$

D none of the above

Question 16
If a circle with equation $x^2 + y^2 = 15$ passes through the point $(k, 2k)$ then k must be equal to:

A 5

B $\sqrt{3}$ or $-\sqrt{3}$

C $\sqrt{5}$ or $-\sqrt{5}$

D 3

Question 17
If P is the point $(2, -1, 0)$, and Q is the point $(-3, 4, 5)$, then the point R which divides PQ in the ratio 2:3 has coordinates:

A $(0, 1, 2)$
B $(-1, 2, 3)$
C $(3, 3, -3)$
D $(0, -1, -2)$

Question 18
In the equilateral triangle shown, each side is of length 4 units. Vectors **a** and **b** are as shown. The scalar product **a** . **b** is equal to:

A 8

B 16

C $4\sqrt{3}$

D 2

Test Set 1 continued

Question 19

The integral $\int_0^{\frac{\pi}{2}} \cos 4\theta\, d\theta$ is equal to:

A $\dfrac{1}{4}$

B 1

C -1

D 0

Question 20

If $\log_p k = 4\log_p v + \log_p 7$, then k expressed in terms of v is:

A $v^4 + 7$

B $4v + 7$

C $28v$

D $7v^4$

Test Set 2

Question 1

The line which makes an angle of 30° with the x-axis and passes through the point $(-1, 2)$ has equation:

A $\quad y - \sqrt{3}x = 2 + \sqrt{3}$
B $\quad \sqrt{3}y - x = 2\sqrt{3} - 1$
C $\quad \sqrt{3}y - x = 1 + 2\sqrt{3}$
D $\quad \sqrt{3}y - x = -2 - \sqrt{3}$

Question 2

If $f(x) = 2x - 2$ and $g(x) = x^2$ then $f(g(x))$ is equal to:

A $\quad 2x^2 - 2$
B $\quad (2x - 2)^2$
C $\quad 4x^2 - 2$
D $\quad 2(x^2 - 2)$

Question 3

Which of the following is most likely to be the equation of the function $g(x)$ in the diagram:

A $\quad g(x) = (x - 5)^2$
B $\quad g(x) = x^2 - 5x$
C $\quad g(x) = x^2 + 5x$
D $\quad g(x) = x^2 - 5$

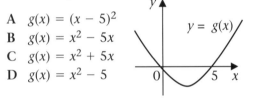

Question 4

The curve with equation $y = \log_3 x - 2$ passes through the point $(3, k)$ when k has the value:

A $\quad -2$
B $\quad 7$
C $\quad -1$
D $\quad 1$

Question 5

The diagram shows the graph of:

A $\quad y = 2\cos(x - 30)°$
B $\quad y = 2\sin(x + 30)°$
C $\quad y = \cos(2x + 30)°$
D $\quad y = 2\cos(x + 30)°$

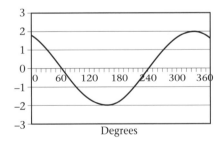

Degrees

Question 6

The exact value of $\cos \dfrac{2\pi}{3}$ is:

A $\quad \dfrac{\sqrt{3}}{2}$

B $\quad \dfrac{1}{2}$

C $\quad -\dfrac{1}{2}$

D $\quad -\dfrac{\sqrt{3}}{2}$

Test Set 2 continued

Question 7

If $y = 2x^3 - 4$ then $\dfrac{dy}{dx}$ at the point

where $x = 2$ is equal to:

A 0
B 24
C 20
D 8

Question 8

If $f(x) = 3x^2 + \dfrac{1}{x}$, $x \neq 0$, then $f'(x)$ is

equal to:

A $6x - \dfrac{2}{x^2}$

B $6x + \dfrac{1}{x^2}$

C $6x - \dfrac{1}{x^2}$

D $x^3 + 1$

Question 9

The graph shows the derived function $g'(x)$ of $g(x)$. For what values of x is $g(x)$ an increasing function:

A $x < 1$
B $x > 1$
C $-1 < x < 3$
D $x < -1$ or $x > 3$

Question 10

If $2x^3 + 5x^2 - x - 1$ is divided by $2x - 1$, the remainder is:

A $\dfrac{1}{2}$

B 1
C 5
D 0

Question 11

If the equation $px^2 - 5x + k = 0$ has real distinct roots then :

A $4pk < 25$
B $4pk > 25$
C $pk < 25$

D $pk < \dfrac{5}{2}$

Question 12

The integral with respect to x of

$3x^2 - \dfrac{3}{x^2}$ is:

A $x^3 + \dfrac{3}{x} + C$

B $x^3 - \dfrac{3}{x} + C$

C $6x + \dfrac{6}{x} + C$

D $3x^3 - \dfrac{3}{x} + C$

continued ➤

Test Set 2 continued

Question 13

If in the diagram the angles α and β are as shown and the lines have the lengths given, then $\sin\angle PQS$ is equal to:

A $\dfrac{63}{65}$

B $\dfrac{56}{65}$

C $\dfrac{99}{65}$

D $\dfrac{33}{65}$

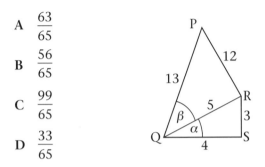

Question 14

The line with equation $y = 3x$ meets the circle with equation $x^2 + y^2 = 160$ at the points:

A $(-4, 12)$ and $(4, -12)$
B $(4, 12)$ and $(-4, -12)$
C $(12, 4)$ and $(-12, -4)$
D $(-12, 4)$ and $(12, -4)$

Question 15

The circle with the centre shown and which passes through the point given has equation:

A $(x - 2)^2 + (y - 3)^2 = 18$
B $(x - 3)^2 + (y - 2)^2 = 18$
C $(x - 3)^2 + (y - 2)^2 = 10$
D $x^2 + (y + 1)^2 = 18$

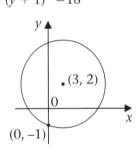

Question 16

If T is the point $(2, -1, 5)$, and V is the point $(-1, 2, 4)$, then vector $\overrightarrow{TV}$ has components:

A $\begin{pmatrix} -3 \\ 1 \\ -1 \end{pmatrix}$

B $\begin{pmatrix} 1 \\ 1 \\ 9 \end{pmatrix}$

C $\begin{pmatrix} 3 \\ -3 \\ 1 \end{pmatrix}$

D $\begin{pmatrix} -3 \\ 3 \\ -1 \end{pmatrix}$

Test Set 2 continued

Question 17

Points A(-1, 3, 4), B(3, 1, -2) and C(5, h, -5) are collinear if h has the value:

A -1

B 5

C 0

D -6

Question 18

If $y = \sqrt[3]{(5x + 2)^2}$ then $\dfrac{dy}{dx}$ is equal to:

A $\dfrac{10}{3\sqrt[3]{(5x + 2)}}$

B $\dfrac{15}{2\sqrt[3]{(5x + 2)}}$

C $\dfrac{2}{3\sqrt[3]{(5x + 2)}}$

D $\dfrac{3}{25\sqrt[3]{(5x + 2)^5}}$

Question 19

If $2\log_k x - \log_k 4 = \log_k 9$ ($x > 0$), then x has the value:

A 36

B 18

C 6

D $\sqrt{13}$

Question 20

When $5\cos\theta + 2\sin\theta$ ($0 \le \theta \le 360$) is written in the form $k\cos(\theta + \alpha)$ ($0 \le \alpha \le 360$), the values of k and $\tan\alpha$ are:

A $k = \sqrt{29}$ $\tan\alpha = \dfrac{2}{5}$

B $k = 7$ $\tan\alpha = -\dfrac{2}{5}$

C $k = \sqrt{29}$ $\tan\alpha = -\dfrac{2}{5}$

D $k = \sqrt{21}$ $\tan\alpha = \dfrac{2}{5}$

Test Set 3

Question 1

The straight line through the point $(4, -1)$ which is parallel to the line with equation $3y = 2x - 1$ has equation:

A $\quad y - 2x = -9$

B $\quad 3y - 2x = -11$

C $\quad 3y - 2x = -10$

D $\quad 3y - 2x = -2$

Question 2

The lines with equations
$2x + 4y + 18 = 0$ and $3x + 2y + 7 = 0$ meet at the point:

A $\quad (-5, 1)$

B $\quad (1, 5)$

C $\quad (-1, -2)$

D $\quad (1, -5)$

Question 3

The equation of the function $f(x)$ in the diagram is most likely to be:

A $\quad f(x) = -x^2 + x + 2$

B $\quad f(x) = x^2 - x - 2$

C $\quad f(x) = -x^2 - x - 2$

D $\quad f(x) = -x^2 - x + 2$

Question 4

The curve with equation $y = 5^x$ meets the y-axis at:

A $\quad (0, 5)$

B $\quad (1, 0)$

C $\quad (0, 0)$

D $\quad (0, 1)$

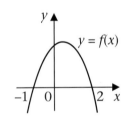

Question 5

The parabola shown in the diagram has an axis of symmetry whose equation is:

A $\quad 2y = b + a$

B $\quad 2x = a - b$

C $\quad 2x = a + b$

D $\quad 2x = b - a$

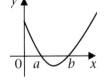

Question 6

The function $k(x) = 2 - 3\cos x$ has a minimum value:

A $\quad -1$

B $\quad 5$

C $\quad 1$

D $\quad 2$

Test Set 3 continued

Question 7

A sequence defined by the recurrence relation $u_{n+1} = 0\cdot6u_n + 300$ has a limit (as n tends to infinity) of:

A 75

B 300

C 500

D 750

Question 8

The curve with equation $y = 3x^2 - 4x + 1$ has a turning point when x has the value:

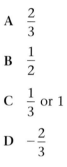

A $\dfrac{2}{3}$

B $\dfrac{1}{2}$

C $\dfrac{1}{3}$ or 1

D $-\dfrac{2}{3}$

Question 9

When $5x^3 + 2x^2 - x - 10$ is divided by $x + 2$, the remainder is:

A 44

B 40

C 36

D −40

Question 10

If the equation $px^2 - 5x + q = 0$ has no real roots then:

A $4pq > 5$

B $4pq > -25$

C $4pq > 25$

D $4pq < 25$

Question 11

The integral $\int (2x + 5)^2 \, dx$ is equal to:

A $\dfrac{1}{3}(2x + 5)^3 + C$

B $\dfrac{4}{3}x^3 + 25x + C$

C $\dfrac{1}{6}(2x + 5)^3 + C$

D $\dfrac{1}{6}(2x + 5) + C$

Question 12

The integral $\displaystyle\int_2^3 3t^2 \, dt$ is equal to:

A 1

B 19

C 21

D 57

continued ➤

Test Set 3 continued

Question 13

If $\tan \alpha = \dfrac{3}{7}$ then $\sin 2\alpha$ is equal to:

A $\dfrac{6}{\sqrt{58}}$

B $\dfrac{42}{\sqrt{58}}$

C $\dfrac{42}{58}$

D $\dfrac{40}{58}$

Question 14

If $\sin \alpha = \dfrac{1}{\sqrt{8}}$, and $\sin \beta = \dfrac{1}{\sqrt{10}}$ then

$\sin (\alpha + \beta)$ is equal to:

A $\dfrac{3 + 3\sqrt{7}}{\sqrt{80}}$

B $\dfrac{3 + \sqrt{7}}{\sqrt{80}}$

C $\dfrac{\sqrt{8} + \sqrt{10}}{\sqrt{80}}$

D $\dfrac{3 - \sqrt{7}}{\sqrt{80}}$

Question 15

The circle with equation
$x^2 + y^2 - 10x + 2y - 10 = 0$ has
centre (p, q) and radius r. The values of
p, q, and r, are:

A $p = 10$, $q = -2$, and $r = \sqrt{113}$
B $p = 5$, $q = -1$, and $r = 4$
C $p = 5$, $q = -1$, and $r = 6$
D $p = -5$ $q = 1$, and $r = 6$

Question 16

The line with equation $y = x$ cuts the
circle with equation
$x^2 + y^2 - 6x - 2y - 24 = 0$ at the
points:

A $(-2, -2)$ and $(6, 6)$
B $(-4, -4)$ and $(6, 6)$
C $(2, 2)$ and $(6, 6)$
D $(4, 4)$ and $(-6, -6)$

Question 17

If points A $(3, -1, 4)$, B $(1, -5, 1)$ and
C $(-1, k, -2)$ are collinear, then k has
the value:

A -9
B -5
C -7
D 9

Question 18

The angles between the vectors **r**, **s**,
and **t** are as shown.

If $|\mathbf{r}| = 2$, $|\mathbf{s}| = 2$, and $|\mathbf{t}| = 3$, then
$\mathbf{s} \cdot (\mathbf{r} + \mathbf{t})$ is equal to:

A $2 + 3\sqrt{3}$

B $\dfrac{4}{\sqrt{2}} + 3\sqrt{3}$

C $2 + \dfrac{6}{\sqrt{2}}$

D $2\sqrt{3} + \dfrac{6}{\sqrt{2}}$

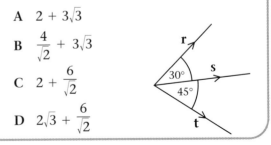

Test Set 3 continued

Question 19

The value of $\log_2 6 + \log_2 4 - \log_2 12$ is:

A 0

B 1

C 2

D 4

Question 20

If $2\cos(x - \frac{\pi}{6}) = \sqrt{3}$ $(0 \leq x \leq 2\pi)$,

then the possible values of x are:

A $\dfrac{\pi}{12}$ or π

B 0 or $\dfrac{\pi}{3}$ or 2π

C $\dfrac{\pi}{6}$ or $\dfrac{5\pi}{6}$

D $\dfrac{\pi}{3}$ or $\dfrac{5\pi}{3}$

Test Set 4

Question 1

In the diagram the line joining points P and Q has equation:

A $2y + 3x = 0$
B $3y + 2x = 5$
C $3y + 2x = 13$
D $3y + 2x = -11$

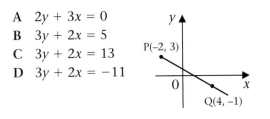

Question 2

If $f(x) = x^3 + 1$ and $g(x) = x - 5$, then $f(g(2))$ is equal to:

A -8
B -26
C 28
D 4

Question 3

If $y^2 + 14y - 2$ is expressed in the form $(y + t)^2 - n$ then t and n have the values:

A $t = 7, n = -51$
B $t = 7, n = 9$
C $t = 7, n = 51$
D $t = 7, n = 2$

Question 4

The unbroken line in the diagram shows the graph of $y = g(x)$. The dotted line shows the graph of:

A $y = -g(x)$
B $y = g(-x)$
C $y = -g(x) - 2a$
D $y = \dfrac{1}{g(x)}$

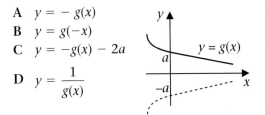

Question 5

If $\sin^2 x° = \dfrac{1}{2}$ $(0 \leq x \leq 360)$, then the value of x must be:

A 45
B 30 or 150
C 45 or 135
D 45 or 135 or 225 or 315

Question 6

If in a recurrence relation
$u_n = 0 \cdot 2u_{n-1} + 50$, and $u_0 = 250$, then u_3 is equal to:

A 2350
B 14·4
C 70
D 64

Test Set 4 continued

Question 7

At the point where $x = 2$, the tangent to the curve $y = 3x^2 - 9x$ has gradient:

A -10
B -6
C 3
D 12

Question 8

The line in the diagram shows the graph of $h'(x)$ where $h'(x)$ is the gradient function of $h(x)$. The function $h(x)$ is an increasing function for:

A $x > -2$
B $x > 0$
C   all values of x
D   no values of x

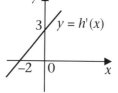

Question 9

When $x^3 + 3x^2 - 4$ is divided by $(x + 2)$ the other factors are:

A $(x - 2)$ and $(x + 1)$
B $(x - 3)$ and $(x + 2)$
C $(x - 1)$ and $(x + 2)$
D $(x + 2)$ and $(x + 1)$

Question 10

The line with equation $y = k$ will not meet the curve with equation $y = x^2 - 4x$ if:

A $k > -4$
B $k < -4$
C $k < 2$
D $k < 4$

Question 11

The integral with respect to x of

$\dfrac{-4}{\sqrt[3]{x^5}}$ is equal to:

A $\dfrac{-20}{3\sqrt[3]{x^6}} + C$

B $\dfrac{8}{3\sqrt[3]{x^2}} + C$

C $\dfrac{3}{2\sqrt[3]{x^8}} + C$

D $\dfrac{6}{\sqrt[3]{x^2}} + C$

Question 12

The integral $\displaystyle\int_{-1}^{1} (3x^2 - 3)\,dx$ is equal to:

A 0
B 4
C -4
D 2

continued ➤

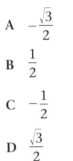

Test Set 4 continued

Question 13

The expression
$\cos 45°\cos 15° + \sin 45°\sin 15°$ is equal to:

A $\quad -\dfrac{\sqrt{3}}{2}$

B $\quad \dfrac{1}{2}$

C $\quad -\dfrac{1}{2}$

D $\quad \dfrac{\sqrt{3}}{2}$

Question 14

The circle with centre $(2, -1)$ and
radius of length 6 has equation:

A $\quad x^2 + y^2 - 4x + 2y - 31 = 0$
B $\quad x^2 + y^2 + 2x - 4y - 31 = 0$
C $\quad x^2 + y^2 - 4x + 2y - 1 = 0$
D $\quad x^2 + y^2 + 4x - 2y - 31 = 0$

Question 15

The tangent at the point $(2, 3)$ to the
circle with equation
$x^2 + y^2 - 6x + 10y - 31 = 0$ has
gradient:

A $\quad -8$

B $\quad -\dfrac{1}{2}$

C $\quad -\dfrac{1}{8}$

D $\quad \dfrac{1}{8}$

Question 16

If $\mathbf{t} = \begin{pmatrix} 1 \\ -4 \\ 2 \end{pmatrix}$ and $\mathbf{v} = \begin{pmatrix} -2 \\ 1 \\ 5 \end{pmatrix}$, then

$2\mathbf{t} - 3\mathbf{v}$ is equal to:

A $\quad \begin{pmatrix} 8 \\ -5 \\ -3 \end{pmatrix}$

B $\quad \begin{pmatrix} 8 \\ -11 \\ -11 \end{pmatrix}$

C $\quad \begin{pmatrix} -4 \\ -5 \\ 19 \end{pmatrix}$

D $\quad \begin{pmatrix} 8 \\ -5 \\ -1 \end{pmatrix}$

Test Set 4 continued

Question 17

If vectors **p** and **q** have components

$\begin{pmatrix} 2 \\ -3 \\ 5 \end{pmatrix}$ and $\begin{pmatrix} 1 \\ 4 \\ k \end{pmatrix}$ and are perpendicular,

then k has the value:

A 3

B 1

C -2

D 2

Question 18

If $y = \cos \dfrac{2}{x}$ $(x \neq 0)$ then $\dfrac{dy}{dx}$ is equal to:

A $-\dfrac{1}{x^2} \sin \dfrac{2}{x}$

B $\dfrac{2}{x^2} \sin \dfrac{2}{x}$

C $-\sin \dfrac{2}{x}$

D $-\dfrac{2}{x} \sin \dfrac{2}{x}$

Question 19

In the diagram $f(x) = \log_k(x - n)$.
From the graph the values of k and n are:

A $k = 5, n = 2$

B $k = 7, n = 3$

C $k = 3, n = 7$

D $k = 2, n = 5$

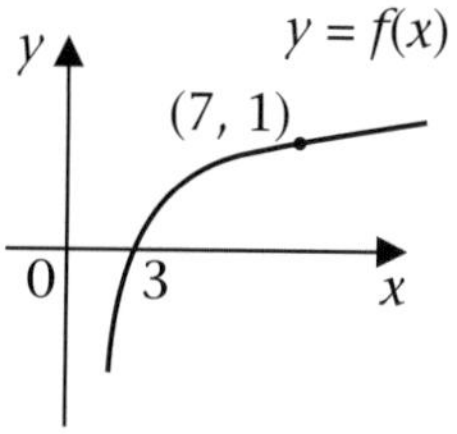

Question 20

When $\sqrt{3} \cos x° + \sin x°$ $(0 \leq x \leq 360)$ is written in the form $k\sin(x - \alpha)°$ $(0 \leq \alpha \leq 360)$, the values of k and $\tan \alpha$ are

A $k = 4$ $\tan \alpha = -\sqrt{3}$

B $k = 2$ $\tan \alpha = \dfrac{1}{\sqrt{3}}$

C $k = 2$ $\tan \alpha = \sqrt{3}$

D $k = 2$ $\tan \alpha = -\sqrt{3}$

Test Set 5

Question 1

In the diagram, points Q (1, 4), S (−2, 1) and R (2, 1) are three vertices of a parallelogram. If P is the fourth vertex, the equation of SP is:

A $3y - 7x = -23$
B $7y - 3x = 13$
C $7y - 3x = 9$
D $7y - 3x = -1$

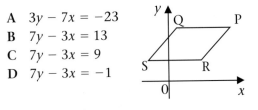

Question 2

If $f(x) = 3 \cos x$, and $g(x) = 2x$, then $g(f(x))$ is equal to:

A $6 \cos x$
B $3 \cos 2x$
C $6 \cos 2x$
D $3 \cos 2x^2$

Question 3

The graph of $y = (x + 8)^2 - 4$ has a turning point (r, s) when r and s have the values:

A $r = -4, s = -8$
B $r = -8, s = 4$
C $r = -8, s = -4$
D $r = 8, s = 4$

Question 4

The parabola shown in the diagram has equation $y = kx(x - a)$. The values of a and k are:

A $a = 4, k = \dfrac{3}{2}$

B $a = 4, k = \dfrac{5}{2}$

C $a = 4, k = -3$

D $a = 4, k = -\dfrac{3}{2}$

Question 5

If $\cos^2 x° = \dfrac{3}{4}$ $(0 \le x \le 360)$, then x must be equal to:

A 60, 120, 240, or 300
B 30, 150, 210, or 330
C 30 or 330
D 60 or 300

Question 6

As n tends to infinity, the sequence defined by the recurrence relation $u_{n+1} = 0{\cdot}7u_n + 21$ has a limit equal to:

A 70
B 30
C 63
D 147

Test Set 5 continued

Question 7

If $y = (2x + 1)(2x - 1)$, then $\dfrac{dy}{dx}$ is:

A   4

B $8x$

C $8x - 1$

D $8x - 4$

Question 8

If $(x - 3)$ is a factor of
$x^3 - 2x^2 - ax + 6$, then a has the value:

A -5

B   5

C   13

D   15

Question 9

If the equation $2x^2 - kx + 5 = 0$ has equal roots, then:

A $k = \pm 4\sqrt{10}$

B $k = \pm 2\sqrt{10}$

C $-2\sqrt{5} < k < 2\sqrt{5}$

D $-2\sqrt{10} < k < 2\sqrt{10}$

Question 10

The line with equation $y = x + 7$ meets the curve with equation $y = x^2 + 2x - 5$ at points whose x coordinates are:

A -2 and -1

B -4 and 3

C -3 and 4

D   2 and 1

Question 11

The integral $\int (x^4 - x^{-\frac{2}{3}} + 5)dx$ is equal to:

A $4x^3 + \dfrac{2}{3}x^{-\frac{5}{3}} + C$

B $\dfrac{1}{5}x^5 + \dfrac{3}{2}x^{\frac{1}{3}} + 5x + C$

C $\dfrac{1}{5}x^5 - 3x^{\frac{1}{3}} + 5x + C$

D $\dfrac{1}{5}x^5 + 3x^{\frac{1}{3}} + 5x + C$

Question 12

The integral $\displaystyle\int_1^2 \dfrac{1}{x^3}\,dx$ is equal to:

A $-\dfrac{5}{8}$

B $\dfrac{3}{8}$

C $-\dfrac{3}{8}$

D $\dfrac{5}{8}$

continued ➤

Test Set 5 continued

Question 13

If $\sin P = \dfrac{3}{5}$ and $\sin Q = \dfrac{5}{13}$ then

$\sin (P + Q)$ is equal to:

A $\quad \dfrac{72}{65}$

B $\quad \dfrac{56}{65}$

C $\quad \dfrac{64}{65}$

D $\quad \dfrac{63}{65}$

Question 14

The expression

$\cos 75°\cos15° - \sin75°\sin15°$

is equal to:

A $\quad 0$

B $\quad \dfrac{1}{2}$

C $\quad 1$

D $\quad \dfrac{\sqrt{3}}{2}$

Question 15

The circle with equation
$x^2 + y^2 - 10x - 4y + 9 = 0$ has centre
(a, b) and radius r. The values of
a, b, and r are:

A $\quad a = 10$, $b = 4$, and $r = 3$
B $\quad a = 5$, $b = 2$, and $r = \sqrt{20}$
C $\quad a = -5$, $b = -2$, and $r = \sqrt{20}$
D $\quad a = 5$, $b = 2$, and $r = \sqrt{38}$

Question 16

If M is the point $(-3, 2, 6)$ and N is
the point $(4, -5, -1)$ and Q divides
MN in the ratio 3:4, then Q has
coordinates:

A $\quad (1, -1, 2)$
B $\quad (1, -2, 2)$
C $\quad (0, -1, 3)$
D $\quad (-1, 2, -2)$

Question 17

Triangle PQR is right angled at R with
sides of lengths shown. The scalar
product $\overrightarrow{QP}.\overrightarrow{QR}$ is equal to:

A $\quad 0$

B $\quad \dfrac{25}{13}$

C $\quad 25$

D $\quad \dfrac{5\sqrt{65}}{13}$

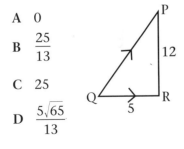

Question 18

The tangent to the curve with
equation $y = \sin 3\theta$ at the point where

$\theta = \dfrac{\pi}{3}$ has gradient:

A $\quad -3$

B $\quad 3$

C $\quad -1$

D $\quad 0$

Test Set 5 continued

Question 19

If the unbroken line in the diagram shows the graph of $y = \log_a x$ the dotted line shows:

A $y = \log_a(-x)$

B $y = -\log_a x$

C $y = a^{-x}$

D $y = \dfrac{1}{\log_a x}$

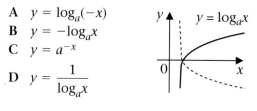

Question 20

If $\cos(x - \dfrac{\pi}{4}) = \dfrac{1}{2}(0 \le x \le 2\pi)$ then

the possible values of x are:

A $\dfrac{7\pi}{12}$

B $\dfrac{7\pi}{12}$ or $\dfrac{23\pi}{12}$

C $\dfrac{5\pi}{12}$ or $\dfrac{13\pi}{12}$

D $\dfrac{\pi}{12}$ or $\dfrac{7\pi}{12}$

Test Set 6

Question 1

If a line has gradient 5 and passes through the points (6, 0) and $(k, -2)$, then k has the value:

A $\dfrac{28}{5}$

B $\dfrac{4}{5}$

C $\dfrac{32}{5}$

D -4

Question 2

When $x^2 - 6x - 10$ is expressed in the form $(x - a)^2 + k$, a and k have the values:

A $a = 3, k = -19$

B $a = 3, k = -1$

C $a = 3, k = -7$

D $a = 6, k = -46$

Question 3

If $f(x) = \sin x$, and $g(x) = x^2 - 1$, then $f(g(x))$ is equal to:

A $\sin^2 x - 1$

B $\sin x^2 - 1$

C $\sin(x^2 - 1)$

D $\sin^2(x^2 - 1)$

Question 4

The unbroken line in the diagram shows the graph of $y = f(x)$. The dotted line shows the graph of :

A $y = f(x - 2)$

B $y = f(x + 2)$

C $y = f(-x)$

D $y = f(x) + 2$

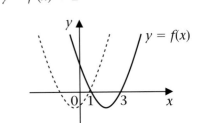

Question 5

If $\cos (x + 30)^\circ = \dfrac{1}{2}$ $(0 \leq x \leq 360)$

then x is equal to:

A 0 or 300

B 30 or 270

C 60 or 300

D 90 or 330

Question 6

If in a recurrence relation $u_{n+1} = 1\cdot5u_n - 3$ and $u_0 = 4$, then u_3 is equal to:

A $2\cdot25$

B $0\cdot75$

C $-0\cdot75$

D $-1\cdot5$

Test Set 6 continued

Question 7

If $g(x) = 6x^3 + 2\sqrt{x}$ then $g'(x)$ is equal to:

A $\quad 18x^2 + \dfrac{1}{\sqrt{x}}$

B $\quad 6x^2 + \dfrac{1}{\sqrt{x}}$

C $\quad 18x^2 - \dfrac{1}{\sqrt{x}}$

D $\quad 18x^2 + \dfrac{2}{\sqrt{x}}$

Question 8

The curve with equation $y = px^2 + qx + c$ has a turning point where x has the value:

A $\quad \dfrac{q}{2p}$

B $\quad -\dfrac{q}{2p}$

C $\quad -\dfrac{2p}{q}$

D $\quad 0$

Question 9

If $x^4 + 3x^3 - kx^2 - 12x + 16$ divides exactly by $(x - 1)$ then k has the value:

A $\quad -8$

B $\quad 8$

C $\quad 11$

D $\quad 26$

Question 10

If $f(x) = (x - 2)(x + 5)$, then $f(x) < 0$ if

A $\quad x < -5$ and > 2

B $\quad x > -5$ and < 2

C $\quad x > 2$

D $\quad x < -5$

Question 11

The integral $\displaystyle\int \dfrac{4 + t^4}{t^3}\,dt$ is equal to:

A $\quad -\dfrac{1}{t^2} + \dfrac{t^2}{2} + C$

B $\quad -\dfrac{12}{t^2} + 1 + C$

C $\quad \dfrac{2}{t^2} + \dfrac{t^2}{2} + C$

D $\quad -\dfrac{2}{t^2} + \dfrac{t^2}{2} + C$

Question 12

If $\displaystyle\int_1^k x^2\,dx = 7$ $(k > 1)$, then k has the value:

A $\quad \sqrt[3]{20}$

B $\quad 2$

C $\quad \sqrt[3]{24}$

D $\quad \sqrt[3]{22}$

continued ➤

Test Set 6 continued

Question 13

If θ is an acute angle such that

$\cos \theta = \dfrac{2}{\sqrt{7}}$ then $\cos 2\theta$ is equal to:

A $-\dfrac{1}{7}$

B $\dfrac{1}{7}$

C $\dfrac{4}{7}$

D $\dfrac{4\sqrt{3}}{7}$

Question 14

The point $(a, 3)$ lies outside the circle with equation $x^2 + y^2 = 100$ if:

A $a^2 > 1$

B $a^2 > 97$

C $a^2 < 91$

D $a^2 > 91$

Question 15

Two circles with equations
$x^2 + y^2 + 4x + 2y - 11 = 0$ and
$x^2 + y^2 - 6x - 4y - 3 = 0$ have
centres P and Q. The distance between
P and Q is:

A 8 units

B $\sqrt{10}$ units

C $\sqrt{136}$ units

D $\sqrt{34}$ units

Question 16

If P is the point $(5, 1, -2)$ and Q is the point $(-1, 3, 2)$, then vector $\overrightarrow{PQ}$ has components:

A $\begin{pmatrix} 4 \\ 4 \\ 0 \end{pmatrix}$

B $\begin{pmatrix} -6 \\ 2 \\ 4 \end{pmatrix}$

C $\begin{pmatrix} -6 \\ 2 \\ 0 \end{pmatrix}$

D $\begin{pmatrix} 6 \\ -2 \\ -4 \end{pmatrix}$

Test Set 6 continued

Question 17

If in terms of unit vectors
$\mathbf{p} = 5\mathbf{i} - 2\mathbf{j} + \mathbf{k}$ and $\mathbf{q} = \mathbf{i} - 4\mathbf{j} - 2\mathbf{k}$,
then $2\mathbf{p} + \mathbf{q}$ has components:

A $\begin{pmatrix} 11 \\ -8 \\ 0 \end{pmatrix}$

B $\begin{pmatrix} 11 \\ -8 \\ -1 \end{pmatrix}$

C $\begin{pmatrix} 10 \\ -8 \\ -2 \end{pmatrix}$

D $\begin{pmatrix} 10 \\ -8 \\ 0 \end{pmatrix}$

Question 18

If $g(x) = \dfrac{4}{\sqrt{(4 - 7x)}}$ then $g'(x)$ is equal to:

A $\dfrac{14}{\sqrt{(4 - 7x)}}$

B $\dfrac{-14}{\sqrt{(4 - 7x)^3}}$

C $\dfrac{-2}{\sqrt{(4 - 7x)^3}}$

D $\dfrac{14}{\sqrt{(4 - 7x)^3}}$

Question 19

If $\log_7 (3x - 5) - \log_7 (x + 2) = \log_7 2$,
then x is equal to:

A -3

B 9

C $4\cdot5$

D $\dfrac{-103}{46}$

Question 20

When $\sin x + 2 \cos x (0 \leq x \leq 360)$ is
expressed in the form $k \sin(x + \alpha)$
$(0 \leq \alpha \leq 360)$, the values of k and $\tan \alpha$ are:

A $k = \sqrt{5}$ $\tan \alpha = 2$

B $k = \sqrt{5}$ $\tan \alpha = \dfrac{1}{2}$

C $k = 5$ $\tan \alpha = 2$

D $k = \sqrt{5}$ $\tan \alpha = -2$

ANSWERS

The difficulty level of each question is shown in brackets.

UNIT 1

Section 1 The straight line

1	D (1)	6	D (2)	11	B (2)	16	A (2)
2	D (1)	7	C (2)	12	A (3)	17	A (3)
3	C (1)	8	B (2)	13	C (2)	18	C (3)
4	B (1)	9	D (2)	14	C (2)	19	A (2)
5	C (2)	10	C (1)	15	C (2)	20	D (3)

Section 2 Functions and graphs of functions

21	B (2)	26	C (3)	31	C (2)	36	D (2)
22	D (2)	27	B (3)	32	B (2)	37	B (2)
23	B (2)	28	A (2)	33	D (4)	38	C (3)
24	B (2)	29	A (3)	34	B (4)		
25	B (2)	30	A (3)	35	C (3)		

Section 3 Trigonometry: graphs and equations

39	D (4)	42	C (4)	45	A (4)	48	B (3)
40	C (4)	43	D (2)	46	C (3)		
41	B (4)	44	A (3)	47	C (2)		

Section 4 Recurrence relations

49	B (3)	52	B (1)	55	B (3)	58	A (2)
50	D (3)	53	D (2)	56	B (3)		
51	B (1)	54	C (4)	57	D (3)		

Section 5 Differentiation

59	C (2)	63	B (3)	67	D (3)	71	B (2)
60	B (2)	64	D (4)	68	D (3)	72	B (2)
61	A (2)	65	B (3)	69	A (3)		
62	A (3)	66	C (3)	70	D (4)		

UNIT 2

Section 1 Polynomials

73	C (1)	77	A (3)	81	C (4)	85	C (2)
74	A (2)	78	D (3)	82	B (4)	86	B (2)
75	B (2)	79	B (3)	83	D (4)		
76	D (3)	80	A (4)	84	A (4)		

Section 2 Quadratic functions

87	B (2)	93	B (2)	99	A (2)	105	C (4)
88	A (3)	94	C (2)	100	D (3)	106	B (4)
89	B (2)	95	A (4)	101	B (2)	107	A (4)
90	B (2)	96	A (4)	102	D (2)	108	D (1)
91	B (2)	97	D (4)	103	D (2)		
92	C (3)	98	A (4)	104	A (3)		

Section 3 Integration

109	C (2)	113	D (3)	117	D (3)	121	C (3)
110	C (2)	114	B (4)	118	A (4)	122	D (4)
111	B (3)	115	C (4)	119	A (4)		
112	A (2)	116	D (4)	120	B (5)		

Section 4 3-D Trigonometry and addition formulae

123	B (2)	127	B (3)	131	C (3)	135	D (5)
124	A (2)	128	A (3)	132	C (5)	136	B (5)
125	D (3)	129	B (3)	133	A (3)	137	A (5)
126	A (4)	130	C (3)	134	D (3)	138	C (5)

Section 5 The circle

139	C (2)	143	D (3)	147	D (2)	151	A (5)
140	D (2)	144	A (3)	148	B (2)	152	B (5)
141	B (2)	145	B (3)	149	D (4)		
142	A (4)	146	C (3)	150	D (5)		

ANSWERS

UNIT 3

Section 1 Vectors

153	A (2)	162	A (3)	171	A (3)	180	B (3)
154	D (2)	163	D (3)	172	C (4)	181	D (2)
155	B (3)	164	B (3)	173	C (3)	182	B (2)
156	A (2)	165	C (3)	174	A (3)	183	A (3)
157	C (2)	166	A (3)	175	D (4)	184	C (2)
158	D (2)	167	C (2)	176	B (4)	185	C (3)
159	A (2)	168	B (4)	177	C (3)	186	A (5)
160	C (3)	169	D (2)	178	A (4)		
161	C (2)	170	A (3)	179	C (2)		

Section 2 Further calculus

187	D (2)	192	D (2)	197	C (3)	202	B (4)
188	B (2)	193	B (3)	198	B (3)	203	A (3)
189	A (3)	194	D (3)	199	D (2)	204	B (3)
190	A (3)	195	A (3)	200	D (3)	205	C (3)
191	A (2)	196	A (4)	201	A (4)	206	D (4)

Section 3 Exponential and logarithmic functions

207	B (1)	210	C (2)	213	D (3)	216	C (4)
208	D (1)	211	A (3)	214	B (3)	217	C (3)
209	A (2)	212	B (3)	215	D (4)	218	A (2)

Section 4 The wave function

219	C (3)	222	B (3)	225	B (4)	228	C (5)
220	B (3)	223	D (4)	226	C (4)	229	C (4)
221	D (3)	224	A (4)	227	A (3)	230	D (4)

Answers to Test Sets

TEST SET 1			
1	D	11	B
2	A	12	B
3	C	13	C
4	B	14	A
5	A	15	B
6	B	16	B
7	C	17	A
8	D	18	A
9	A	19	D
10	B	20	D

TEST SET 2			
1	C	11	A
2	A	12	A
3	B	13	A
4	C	14	B
5	D	15	B
6	C	16	D
7	B	17	C
8	C	18	A
9	C	19	C
10	D	20	C

TEST SET 3			
1	B	11	C
2	D	12	B
3	A	13	C
4	D	14	B
5	C	15	C
6	A	16	A
7	D	17	A
8	A	18	D
9	D	19	B
10	C	20	B

TEST SET 4			
1	B	11	D
2	B	12	C
3	C	13	D
4	A	14	A
5	D	15	D
6	D	16	B
7	C	17	D
8	A	18	B
9	C	19	A
10	B	20	D

TEST SET 5			
1	B	11	C
2	A	12	B
3	C	13	B
4	D	14	A
5	B	15	B
6	A	16	C
7	B	17	C
8	B	18	A
9	B	19	B
10	B	20	B

TEST SET 6			
1	A	11	D
2	A	12	D
3	C	13	B
4	B	14	D
5	B	15	D
6	C	16	B
7	A	17	A
8	B	18	D
9	B	19	B
10	B	20	A